EVERYDAY ADVENTURES WITH UNRULY DATA

EVERYDAY ADVENTURES WITH UNRULY DATA

MELANIE FEINBERG

The MIT Press
Cambridge, Massachusetts
London, England

© 2022 Massachusetts Institute of Technology

All rights reserved. No part of this book may be reproduced in any form by any electronic or mechanical means (including photocopying, recording, or information storage and retrieval) without permission in writing from the publisher.

The MIT Press would like to thank the anonymous peer reviewers who provided comments on drafts of this book. The generous work of academic experts is essential for establishing the authority and quality of our publications. We acknowledge with gratitude the contributions of these otherwise uncredited readers.

This book was set in Bembo Book MT Pro by Westchester Publishing Services. Printed and bound in the United States of America.

Library of Congress Cataloging-in-Publication Data
Names: Feinberg, Melanie, 1970- author.
Title: Everyday adventures with unruly data / Melanie Feinberg.
Description: Cambridge, Massachusetts : The MIT Press, [2022] | Includes bibliographical references and index.
Identifiers: LCCN 2021057608 | ISBN 9780262544405 (paperback)
Subjects: LCSH: Information technology—Social aspects. | Knowledge management. | Information organization. | Information society.
Classification: LCC HM851 .F45 2022 | DDC 303.48/33—dc23/eng/20211230
LC record available at https://lccn.loc.gov/2021057608

10 9 8 7 6 5 4 3 2 1

I see a preoccupation with the deep structure, the underlying structure, with the gesso underpainting that is red earth, black earth.
—Gloria Anzaldúa, *Borderlands/La Frontera: The New Mestiza*

CONTENTS

Acknowledgments xi

INTRODUCTION: A SPIRIT OF ADVENTURE 1
   Wayfaring and Transport 1
   Data in the Everyday 2
   Chapter Structure: Adventure and Reflection 4
   Thematic Sequence and Connecting Arguments 5
   This Book as Methodological Invitation 7
   Antecedents 9

1   SERENDIPITY 13
   Adventure: A Trip to the Library 13
      An Uncomfortable Encounter with an Enthusiastic Librarian 13
      An Enchantingly Strange Museum 16
      A Call Number That I Can't Find 18
      Danish Supermarkets 20
      Two Serendipitous Books 24
   Reflection: How Do We Attend to Data in the Everyday? 26
      The Intellectual Heritage of Knowledge Organization 26
      Reciprocal Themes across Multiple Disciplines 28
      Information Design, Power, and Accountability 29

2   OBJECTIVITY 33
   Adventure: That Scale Must Be Wrong! 33
      Stepping on a Scale 33
      Step Counters on Smartphones 35
      The Allure of Numbers 38
      Measurement and Convention 40
   Reflection: How Is the Quantitative Not? 45
      The Two Cultures of the Sciences and the Humanities 45

        Human Judgment in Quantitative Measurement   47
        The Conduit Metaphor and the Mathematical Theory of Communication   52
        Objectivity as That Which Speaks for Itself   54
        The Human Work of Data Collection   57

3   EQUIVALENCE   61
    Adventure: When Is Butter?   61
        Lurpak Butter in Denmark and Lurpak Butter in the United States   61
        Thin Mints in California and Thin Mints in Texas   64
        American Coke and Mexican Coke   65
        Editions of *Hamlet* in North American Library Catalogs   68
        Editions of *Mr Jelly's Business* in the Austlit Catalog   74
    Reflection: Is This Thing the Same as That One?   81
        What *One Thing* Might Constitute   81
        Disciplinary Assumptions about Information Things (Documents)   83
        Decisions about Things in Data Design and Implementation   87

4   INTEROPERABILITY   95
    Adventure: I Can't Cook Rice in Copenhagen   95
        Cooking Rice   95
        French Bread in Different Languages   99
        The Provenance of an Afternoon Pastry   103
        A Thai Robot Curry Taster   105
        Protocols for Removing Dangerous Posts from Social Media   111
    Reflection: How Does the Same Process Lead to Different Outcomes?   114
        Scientific Reproducibility as an Unsettled Concept   114
        The Challenge of Semantic Data Interoperability   116
        Empirical Outcomes in the Pursuit of Interoperable Data   122
        The Imperative of Data Criticism   126

5   TAXONOMY   129
    Adventure: Not That Hierarchy   129
        A Vile Hierarchy   129
        The Vile Hierarchy, Revised   135
        That Jumble Drawer in Your Kitchen   140
        Confessions of a Data Snob   141
        "User-Centered" Taxonomy   145

Reflection: How Does Classificatory Structure Matter? 147
   A Constellation of Fundamental Data Concepts 147
   Taxonomies and Data Users 148
   Taxonomy Development as Design Rather Than Discovery 150
   Structural Constraint as an Invitation to Creativity 154

6 LABELS 159

Adventure: A Mistake on My Mastercard 159
   A Mistaken Name That I Like 159
   Greece and Macedonia 163
   Mrs., Miss, and Ms. 166
   A Magic Wand 169
   Orientals, the Yellow Peril, and the Chinese Virus 170
Reflection: What's in a Name? 176
   Common-Sense Distinctions between Concepts and Labels 176
   Distinctions between Concepts and Labels in the Practical Literature of Controlled Vocabularies 178
   Empirical Realities of Extracting Concepts from Documents 182
   Reconfiguring Concept-Label Relationships 186

7 LOCALITY 191

Adventure: Have You Ever Been to Louisiana? 191
   A Poet Talking about American Racism 191
   Height Measurement at the Doctor's Office 193
   Race and Ethnicity in the US Census 196
   Race and Ethnicity in Online Dating Sites 198
   A Novelist Talking about Danish Racism 207
Reflection: How Is Data Situated? 210
   Constraint and Creativity in Cognitive Categorization 210
   Local Conditions and Cognitive Prototypes 213
   Historical Efforts to Standardize Data Collection across Locations 216
   Data Creation as an Assertion of Values (An Extended Example from Library Cataloging Data) 219
   Cherishing the Humanity in Our Data 225

CONCLUSION: STILL LIFE WITH DATA 231

Notes 243
References 275
Index 301

ACKNOWLEDGMENTS

This book was made possible by funding from the European Union's Horizon 2020 research and innovation program under the Marie Skłodowska-Curie grant agreement No. 793340. Obtaining that crucial fellowship required a lot of help! Many thanks to my longstanding mentor Jens-Erik Mai; the research support staff at the University of Copenhagen (KU) Faculty of Humanities: Tatjana Crgnorac, Mille Møllegaard, and Hans-Christian Køie Poulsen; the Summer Writing Group program sponsored by the Center for Faculty Excellence at the University of North Carolina at Chapel Hill (UNC), and my SWG crew: Alexa Chew, Sarah Dempsey, Ben Frey, and Vaughn Upshaw; and Gary Marchionini, dean of the School of Information and Library Science at UNC.

For giving me the courage to write stories about butter and mayonnaise rather than detailed readings of stultifying datasets, I thank the fall 2018 fellows at the Institute for the Arts and Humanities (IAH) at UNC: Carolina Sá Carvalho, Seth Kotch, Jacqueline Lawton, Tim Marr, Michelle Robinson, William Sturkey, Daniel Wallace, and Lyneise Williams.

For helping to make my time in Copenhagen so productive and pleasant, thank you to everyone at the KU Department of Information Studies and Department of Communication. (Because another lockdown was instituted in Denmark shortly before I left in December 2020, I wasn't able to bring cake to the office on the occasion of my departure; so, I owe you all some cake!) I am especially grateful to have been able to participate in a variety of research groups at KU: the Digital Information and Communication (DICO) research section (coordinated by Laura Skouvig and Klaus Bruns Jensen); the Internet and Society group (coordinated by Sille Obelitz Søe and Rikke Frank Jørgensen), and the Digital Cultures group (coordinated by Bjarki Valtýsson and Kristin Veel). I also benefitted from presenting in-progress versions of this work to the Technologies in Practice salon at the IT University in Copenhagen, to the research seminar at the Department of Arts and Cultural Sciences at Lund University, and to students in Vivien Petras's and Elke Greifeneder's classes at Humboldt University in Berlin.

For time and attention in reading drafts, I am grateful to Amelia Acker, Jack Andersen, Julia Bullard, Tanya Clement, Jonathan Furner, Ayse Gursoy, Heather Houser, Jens-Erik Mai, Daniela Rosner, Irina Shklovski, Sille Obelitz Søe, Olof Sundin, Deborah Turner, and the ORG group at UNC—in particular, stalwarts Ryan Shaw, Neal Thomas, Evan Donohue, Patrick Golden, Elliott Hauser, and Colin Post, along with all the occasional attendees and mailing list lurkers. At a key point, Melissa Feinberg and Paul Hanebrink gave much-needed advice on my book proposal, rescuing me from a dark place. Additional thanks to everyone who responded to the last-minute poll of potential book titles!

For maintaining a sustaining, collegial environment despite challenging times for our institution, I thank the faculty, staff, and students at the School of Information and Library Science at UNC. Academia can be tough in so many ways, and I'm glad we're in it together.

Special thanks to everyone at MIT Press, especially my acquisitions editor, Gita Devi Manaktala; acquisitions assistants Erika Barrios and Suraiya Jetha; production editor Cheryl Hirsch; and copy editor Virginia Schaefer. Erika Millen contributed the fabulous index. Your professionalism, skill, and kindness is unmatched!

When I tell people that I'm a professor, they often associate that with teaching. But doctoral programs train us to be researchers, not teachers. For me, at least, teaching is hard work. It makes me nervous, and I'm not very good at it! My struggles in the classroom, though, have consistently propelled my scholarship, often in ways that I would never have expected. I am especially grateful for my students, who motivate me to reconsider everything I thought I knew already.

Likewise, I owe a tremendous debt to all my teachers: from my first teachers, my parents, through grade school, high school, college, and graduate school. In particular, I want to recognize my doctoral adviser, Allyson Carlyle, a remarkable and compassionate person who passed away unexpectedly in 2020.

Finally and most of all, thank you to Jason, who was there every step of the way.

INTRODUCTION: A SPIRIT OF ADVENTURE

WAYFARING AND TRANSPORT

I am a professor at the University of North Carolina at Chapel Hill, but this book was written during a research fellowship in Copenhagen, Denmark, from June 2019 to December 2020. In Copenhagen, cycling is the predominant way of getting around. Parents regularly bundle two, three, or even four small children into cargo bikes. There are no automobile-based ride-sharing companies like Uber or Lyft in Copenhagen, but there is a popular food-delivery service that operates on a similar model—via bicycle, of course.

I tried riding a bicycle in Copenhagen, but I found it *too* direct. The bike lane felt like an invisible tunnel, disconnected from the surrounding neighborhood. With all the cycle traffic, I had to focus intently on the act of getting to my destination, on the act of cycling itself. I couldn't pause or look around, and if I found myself just a little bit lost, it was hard to recover.

For me, as a stranger to the city, bicycling in Copenhagen was *transport*. In transport mode, according to the anthropologist Tim Ingold, travel is merely something to endure; it's arrival that's important (Ingold 2007). In contrast, Ingold posits, the *wayfaring* mode of travel emphasizes the experience of the journey. *Wayfaring* is about being *in* a place as you move through the landscape, rather than just going *across* an undifferentiated space. In the act of wayfaring, movement is sensemaking: becoming aware of where you are, where you are going, and where you have been. Wayfarers, according to Ingold, are always moving along, and yet also always present, wherever they happen to be.

I felt I would be missing something, if I spent a year in Copenhagen and understood the city only as sequence of separate destinations. To really grasp and appreciate this place, I'd need to see it as a meshwork of connections, where one street merges into another and each route contributes to a larger understanding of the whole. Instead of transporting myself on a bicycle, I wanted *adventures* in my daily commute, even if they were quotidian; adventures that would help me to *inhabit* this new place, in the manner of wayfaring. So I walked in Copenhagen, rather than riding a bike.

It took longer that way. Paradoxically, though, the constraints of my feet gave me a certain freedom—I was able to cultivate a sense of where I was, even when I wasn't quite sure of my precise location, so that I was never really lost. I could be curious as I walked. I could wander a bit. And so, through my rambles, I did come to know the city, and especially my neighborhood: that the busiest times at the fishmonger were Friday evening and Saturday morning, so if I wanted some of that freshly smoked mackerel, I should get it on Thursday; how even in rule-following Denmark, people would constantly leave old clothes on the stoop of the Frelsens Hær (Salvation Army) store, no matter what kind of sign was on the door to tell people not to do that; that the evening-shift cashier at the al-Basra market (named for a city in Iraq) was actually Turkish, and that "Middle Eastern" Danes came from many specific places—Syria, Bosnia, Iraq, Iran. These observations connected into a sense of character, an uncool neighborhood that felt homey once you got to know it, where old and new quietly coexisted but didn't always mix. My understanding of this character, importantly, did not arise merely in visiting different locations but in going from place to place, from the ongoing journey rather than the destinations.

The essays in this book, written during that stay in Copenhagen, are written to emulate wayfaring rather than transport. They are the everyday adventures of a conceptual place rather than a physical one: the landscape of data.

DATA IN THE EVERYDAY

Over the last decade, popular rhetoric around data has given it quasi-divine status: ubiquitous, omnipotent, ineffable. In this incarnation, data has both tremendous power—to predict the future, solve our most pressing concerns, and make us wealthy beyond our wildest dreams—and mysterious inaccessibility. Only a caste of initiates (data scientists) can cause this data to speak, and only with the mediation of arcane spells (algorithms), which are equally unknowable. Having good data, moreover, is seen as vitally important to a smoothly functioning society. The soundness of scientific knowledge and of policies that might derive from that evidence depends on data that is accurate, stable, and certain.

But when data tumbles out of the cloud and into everyday life, it becomes simple and unremarkable. Data, after all, is just a description of a thing's qualities: a summary of a book's plot, the price of a carton of milk, customer reviews of a neighborhood restaurant, the breed of a dog. What's your name? That's data. Where do you live? That's data too. What's your favorite memory from childhood? Also data. Sometimes data is structured in a certain way—not just freeform like a story—but that's familiar also. Window shopping on the Web? Browsing for new attire on a favorite fashion site is a scan of a structured dataset, with sizes and colors expressed as a controlled vocabulary of standardized values: that frock is available in Navy, Smoke, or Moss, in sizes 2–16. Nothing could be more mundane.

These everyday encounters with data, moreover, are often ambiguous—annoyingly and familiarly so. How many of us have struggled with situations like these?

- Spending too much time researching kitchen appliances, because you can't figure out whether Blender A (in black) and Blender B (in stainless steel) are merely the same model of blender in different colors or are different models with slightly different features.
- Looking at your electronic health record and realizing that you are officially five pounds heavier than you think you are, because the doctor's office always weighs you with clothes and shoes.
- Not knowing how to fill out standardized forms because available boxes for race, gender, or marital status don't sufficiently describe your identity.

In this book, I relate adventures with everyday data, writing about what I encounter and how I came to understand it that way. My understanding of these data adventures arises from my expertise in a relatively obscure academic discipline. I'm an information scientist of a particular sort: a classificationist.

As a classificationist, I study the selection, description, and arrangement of collections. The term *collections* often brings to mind physical things, like collections of stamps or paperweights, and some collections do bring together physical things. For example, libraries are collections of books and other recorded information; supermarkets are collections of food and other everyday items; an herb garden is a collection of plants that are used to flavor food and drink. But other collections bring together digital things: streaming video services are collections of time-based visual media, available to watch online; the World Wide Web is a collection of documents encoded in HTML; an image search service is a collection of digital pictures. And sometimes, the *things* being collected are data, in physical or digital form. A library is a collection of books, and a library catalog is a collection of data about books. A library catalog is a dataset. In primary school, when I first started using the library, the catalog of book data was instantiated on paper cards in a special card cabinet. Today the catalog of book data is available online in digital form. Both the card catalog and the online catalog are datasets. A shopping site like Amazon is just another kind of online catalog—another kind of dataset. Indeed, Amazon began as a dataset about books, before it expanded to include data about yoga mats, laptop computers, barbecue grills, and just about everything else under the sun. A social media site like Facebook is also a kind of dataset; a dataset of user profiles.

Another way that I can describe myself as a scholar, then, is like this: I study the design and implementation of datasets. When I publish papers in the field of human-computer interaction, for example, I describe my work this way, because *the design and implementation of datasets* is terminology that computer scientists recognize. But to me as a classificationist, a dataset is just another kind of collection. This is important: it means that an adventure in a library is very similar to an adventure with a library catalog. *If we tell a story about a trip to the library, we're also telling a story about data.*

This premise—that there is a fundamental similarity between collections of physical things, collections of digital things, and collections of data about things—underlies the network of stories that make up this book. My adventures often begin with stories about physical things (like the supermarket). As I travel onward, I may observe conceptual similarities with digital things (like Twitter) and data about things (like step counters on smartphones). One path leads to another.

CHAPTER STRUCTURE: ADVENTURE AND REFLECTION

Besides this introduction and a brief conclusion, the book comprises seven chapters. Each chapter has two parts:

- A main essay, the *adventure*.
- A companion essay, the *reflection*.

The *adventure* essay is a personal account of how I experienced and came to comprehend a set of encounters with everyday data. In these adventures, I both observe and explain, considering and reconsidering the events as I go along. The value of these adventures arises not from the particular details of what I see and experience but from the mode of seeing that I exemplify. In the adventure portion of the chapter, I begin with an anecdote from everyday life, such as mentally arguing with the numbers that appear on a scale (chapter 2) or laughing at a misspelling of my name (chapter 6). Many of these stories arise from the daily happenings of my sabbatical in Copenhagen, and they will probably seem humdrum and familiar. In offering these events as adventures, I invite the reader to consider their own personal experiences, analogous and not, through a similar lens. Accordingly, although these adventure essays incorporate lots of ideas, they don't include many citations to academic scholarship; that's the job of the accompanying reflections.

The *reflection* essay functions as a kind of bibliographical autobiography, extending and amplifying the themes from the adventure via my personal engagement with the scholarship of information science. These reflections synthesize the scholarly substrate that enables me to experience the adventures as I do. Honestly, information science can be hard to dip into. Its concepts tend to be sunk within complicated institutional contexts and obscure legacy technologies. It can seem pretty boring. But it's also a discipline whose practitioners have been thinking about what we now call *data* for decades upon decades—reaching back to the nineteenth century. Through each reflection, I show how the scholarly traditions of information science can offer an intellectually precise, and yet fundamentally humane, perspective on today's data-related concerns.

The two essays that make up each chapter are complementary—the reflection essay provides a more scholarly understanding of the adventure essay's shape and trajectory—but they don't require each other. The adventure essay, in particular, can be read alone. Similarly, the chapters deepen and reinforce each other, but each is independent, and the book need not be read linearly.

THEMATIC SEQUENCE AND CONNECTING ARGUMENTS

Each chapter centers around a primary theme.

Chapter 1 begins with a trip to the library. A welcoming librarian at the University of Copenhagen suggests that I might be inspired by browsing the shelves, and I consider her offer. This chapter considers the value of *serendipity*—of understanding data in the same way that we might understand conventionally expressive forms of communication such as poetry or music.

In chapter 2, I question the validity of numbers on a scale as my weight goes up and down. If I can feel my pants getting tighter and looser, why do I care about those numbers? This chapter reflects on the notion of *objectivity*, especially as it relates to quantitative data.

In chapter 3, I realize that the Lurpak butter we've been buying in Copenhagen is slightly smaller than the same butter as packaged in the United States. This chapter ponders what we mean when we use terms like *the same butter*. I consider the functional *equivalence* of copies and versions, focusing on the uncertainty of reference that permeates most data—what is that thing that we're describing, anyway?

Chapter 4 relates my many failures in attempting to cook rice in our Copenhagen apartment, using this chronicle to examine what standardized replicable processes and standardized interoperable data can really achieve. This chapter provides a critical assessment of *interoperability*, the notion that underlies our ability to aggregate data from multiple sources and be confident in our subsequent use of it.

Chapter 5 describes an ongoing misunderstanding between me and my students regarding the concept of hierarchy. This chapter explains the difference between taxonomic hierarchy and power hierarchy, and advocates for *taxonomic structure* as a logically transparent system of relations.

In chapter 6, I laugh when I see that my new corporate Mastercard has been issued to *Meanie* instead of *Melanie*. This chapter probes the data that inheres within names and considers the functional association between *labels* and concepts in relation to data creation and use.

Chapter 7 tells the story of hearing a Black American poet speak at a Danish art museum, reflecting on the role of time and place in data implementation. This chapter considers the role of *locality* and local practices of data collection in shaping our understanding of the world. Even something as innocuous as units of measurement, this chapter suggests, can invoke a set of unique data practices that in turn affect the concept our data is describing. These situated practices, this chapter contends, contribute to rather than detract from the value of the associated data.

Chapters 3, 5, and 7 have a more methodological bent. These adventures are slightly more technical, in that they demonstrate a process of working through examples in some depth. In contrast, chapters 2, 4, and 6 are more overtly critical, proposing a reconsideration of their destinations. Chapters 3, 4, 5, and 6 circumnavigate a set of concepts—equivalence, interoperability, taxonomy, and labels—that classificationists such as myself have traditionally concerned ourselves with. Chapters 1, 2, and 7 complement those with ideas—serendipity, objectivity, and locality—that I personally find inextricable from the classificationist's purview.

Although each chapter recounts a different set of adventures, there are many points of overlap. Some of these common arguments include the following:

- Data—whether collected and processed by machines or collected and processed by people—is the product of many acts of human interpretation. No system, manual or automated, can fully contain the dynamic creativity of all these aggregated data creators. Data, therefore, will always retain some ambiguity.
- By paying attention to how data works structurally, we can better understand this ambiguity: where it arises, how it manifests, and what we might want to do about it.
- We can sometimes constrain data ambiguity. These efforts always fail in some respects, and time undoes them, but we may want to employ them nonetheless.
- Like poetry or music or any other form of human communication, data is reinterpreted—in a sense, remade—every time that it is used. Our engagement with its ambiguity is therefore an ongoing process. Data is a story that we read and reread; it is not an equation to be solved.

Underlying these arguments is a conviction that data and our comprehension of it arises from a situated, limited perspective on the world. All data bears traces of the history, culture, social relations, and personal circumstances in which it is made. Data, in other words, is an amalgamation of human particulars. To ascribe universality or neutrality to data is therefore both impossible and dangerous.

THIS BOOK AS METHODOLOGICAL INVITATION

My orientation toward data is broadly grounded in feminist standpoint epistemology.[1] In line with these commitments, this is a purposefully personal book. I recount unique experiences—both of everyday events and of the scholarly literature—that arise from my own point of view and set of circumstances. I provide these details for transparency: they document my interpretive process from initial observation to an understanding of that observation's significance. As I describe these evolving reflections, I aim to be interrogative as well as expository, with an eye toward surfacing my interpretive rationale and, particularly, my hesitations, doubts, and reassessments. My goal is not to make simple pronouncements but to share a practice—not a "best practice" to be cut loose from its circumstances and adopted as a general procedure, but a living, empirically situated practice, an actual example of applying a particular form of attention toward a specific set of phenomena. I hope to provide a generative methodological demonstration for the reader to contemplate and adapt, rather than a prescriptive checklist for the reader to copy.

In taking this approach, I attempt a tricky balance. To inspire recognition of analogous experiences, I've selected examples that I hope will seem broadly familiar. Ideally, these stories will prompt the reader to consider similar situations and observations. To encourage this introspection, I sometimes refer to a shared social context. For instance, I might talk about going to the supermarket or looking at step counts on a smartphone as a common thing. At the same time, however, I need to avoid the omniscient, universalizing perspective that I reject, by clarifying my position within the narrative. The supermarket might be a common place, but the way that I engage with the supermarket is much more particular, and it is important for me to be reflexive about that.

Maintaining this balance has been challenging. It has been hard to keep from overgeneralizing my own perspective, which is one of relative privilege. I'm a tenured White professor, a product of elite American universities and currently employed at one. The infrastructure of the so-called meritocracy was designed for people like me to succeed, and I've had many years of accumulated benefit from that. All this gives me a form of power that I carry with me everywhere, perhaps most of all in the everyday situations that I discuss in this book. Likewise, I'm a middle-aged woman who sometimes finds herself crankily ranting about these kids today, underestimating societal changes that I've had the freedom to ignore. (It's been easy for me to avoid social media, for instance, and accordingly to dismiss its power.) In sum: even when I'm gesturing toward a shared context, I am speaking from a place of advantage. Although I've tried to acknowledge that in the text, there are likely places where I have not sufficiently done so.

ANTECEDENTS

In the summer of 2009, I made an unanticipated trip home to Long Beach, California, to help my great aunt Katy, who had responded poorly to chemotherapy. In our family, only my twin sister Melissa and I knew about Katy's cancer. Years prior, Katy had asked us to be her executors, but she had talked vaguely about her plans and she didn't have a will. On that trip to California, one of my tasks was to better understand her wishes so that I could find an attorney and get paperwork drawn up. Concurrently, we both struggled to comprehend her prognosis and treatment options, as I was enjoined to avoid all other relatives. And did I mention that her car had failed to start when I attempted to drive it from the hospital parking lot where it had been sitting? I had to get that fixed, too. . . . Amid these pressures, I yearned for clear instructions and a simple checklist of things to do. But Katy wouldn't provide that for me. I would ask about the attending doctor's visit or about what kinds of organizations she wanted to support in her will, and she would abandon that discussion to banter with the nurses, the technicians, the orderlies, everyone who came into the room—somehow she knew all their names and histories, their dreams and aspirations. Then she would ask me to go to her preferred bakery to order a cake for the hospital staff, and a fruit arrangement as well. Um, Katy . . . the oncologist? The lawyer? My stress intensified. Katy looked at me, my eyes spinning with anxiety, and said I needed to slow down. I needed to trust her. She was showing me *how* she thought, so that I could better understand *what* she thought. The banter and the cake were not extraneous. They were the whole point of what I needed to do.

Also that summer, I was supervising an independent study for Kristen Hogan, a master's student.[2] Kristen was an aspiring librarian, an activist for information equity, and a scholar who had written a dissertation on women's bookstores in North America. Kristen, who wanted to make change in libraries, found the literature of information science lifeless. It was challenging, I agreed, to extract worthwhile ideas from the gunk—dense jargon, legacy technologies, ingrained institutional habits—that encrusted them. But that wasn't what she meant. For Kristen, the manner of the writing was even more problematic: the quasi-scientific structure, the passive voice, the dead sentences. The *how* contaminated the *what*. Kristen didn't say that information science was effectively dead to the rest of the academic world. But I thought a lot, after that, about the barriers my field had erected around itself. I realized that I, too, had contributed to the parochialism of my discipline, probably more than I would like to admit. But how to break that pattern?

The poet and feminist theorist Gloria Anzaldúa was a touchstone for Kristen, and, following our conversations, I reread her 1987 masterwork *Borderlands/La Frontera*. It was a disorienting experience. Anzaldúa's book seemed to exist on a separate plane from my scholarly life, even as I called myself a pluralist in matters of classification theory. Reading *Borderlands* was like a weird echo of being in Katy's hospital room, alienating and exhilarating at the same time. It's not easy, following Gloria Anzaldúa through encounters with the serpent goddess Coatlicue, being transformed into something that we don't really have words for. Even as I found myself intellectually reinvigorated in contemplating the idea of mestiza consciousness as a kind of classificatory warrant, I had no idea what that might mean, not only *technically*, as a matter of data structure, but *rhetorically*, in terms of how to write about it. But in reading Anzaldúa, a conviction started to form: to demonstrate the continuing relevance of information science, the manner of writing would not be extraneous; it would be the whole point of what I needed to do.

That summer, in 2009, it was sometimes really uncomfortable, being shown how someone else thinks, rather than being told what to do. I got frustrated sometimes—in the hospital with Katy, reading Anzaldúa. I wondered whether it would all make sense. Some of it I'm not sure I understood in the way that either of them would have wanted me to. But they set me on a path, both of them.

Katy died the next spring. As she wanted, the wording in the trust that my sister and I set up for her was purposefully vague, which was scary. Ultimately, though, we established a fund in her name with a community foundation, to support programs "that nurture body, mind, and spirit of goal-oriented people, especially youth, in the City of Long Beach." This was the way that we encapsulated Katy's mindset: the person who spent her recovery period learning about the hopes and ambitions of the hospital staff, telling them jokes and getting them cake. And at the moment when we actually wrote the check to the foundation, that December, after we had sold the house and its contents and settled everything that needed to be settled—the sun burst out of a cloudy sky and flooded the conference room with light. That really happened!

Introduction

Expanding the purview of information science—showing its value more broadly—has been a longer road. Omens have not been forthcoming. But this book continues what began that summer, telling ideas through the story of their process in concrete, immediate ways.[3] Unlike Gloria Anzaldúa, my stories are simple and familiar, having to do with Girl Scout cookies and online dating profiles rather than with the goddess Coatlicue. But the destination might be hazy at the beginning of an adventure, and it might be a little frustrating, not being told what to do, and not being able to skim for the important bits, the ones where I lay out my grand argument and explain how novel it is and how I am such a ground-breaking scholar. It's not that I don't have opinions or make arguments in the book—I have both opinions and arguments—but it's the connections that are important, not the destinations. It's how we get there that's the point of it all.

1
SERENDIPITY

The selection, description, and arrangement of any collection—a library, a supermarket, a video streaming service, a database of terrorist events—structures a kind of story about the collection's contents. When that data story reveals an unexpected insight, it's a kind of serendipity. But unfamiliar information systems can be exasperating as well as marvelous, and this chapter explores those contradictions.

In the adventure essay, I tell stories about

- An uncomfortable encounter with an enthusiastic librarian.
- An enchantingly strange museum.
- A call number that I can't find.
- Danish supermarkets.
- Two serendipitous books.

In the reflection essay, I ruminate over

- The intellectual heritage of knowledge organization.
- Reciprocal themes across multiple disciplines.
- Information design, power, and accountability.

ADVENTURE: A TRIP TO THE LIBRARY

An Uncomfortable Encounter with an Enthusiastic Librarian

One summer day, I was bounding down the stairs in the department library, hoping to retrieve some books I had ordered. I was hurrying. The library was staffed only from 10 to 12 each morning, and I didn't want to miss my window.

It was July of 2019, and I had recently arrived at the Department of Information Studies at the University of Copenhagen, where I would spend a year-long sabbatical. Finally, I could begin the book project that I had long contemplated! But my initial elation was giving way to anxiety. After four years of planning and preparation for this trip, I was finding it hard to get started. That's why I had ordered the library books: to refocus my mind on a full page rather than my own blank screen.

I arrived in plenty of time, but when I located the circulation desk, it was empty. As I looked around, I noticed an office behind the desk; its door was ajar. I hesitated, but I really wanted those books. I crossed the threshold of the open doorway, peeking inside. Two women were sitting in front of their computers, chatting to each other. They stopped and looked up at me. Awkwardly, I explained my purpose. No one said anything for a moment, and I began to think I had done something wrong. Slowly, one of the women introduced herself as Karen, the department librarian. Then she waited.

I introduced myself also, feeling sheepish. I hadn't anticipated this; to me, picking up a hold was a quick, anonymous transaction. Upon hearing my name, Karen realized that I was the new visiting scholar. Her mien changed. She rose and began to show me around the library, inquiring solicitously after my research interests. I felt even more confused, unsure of what was expected of me. As I stammered noncommittally, Karen nodded encouragingly. "You might be interested in this area," she suggested, gesturing to a section of a nearby rack. The tour continued.

Her voice rising with enthusiasm, Karen shared her conviction that browsing library shelves might provide un-looked-for inspiration—serendipity—to help me shape my project. I assured Karen that I, too, believed in this kind of serendipity. But my mind was on the books that I was there to retrieve. When it seemed appropriate, I made polite excuses and returned to my office.

Later, as I tried once again to get going with my book ideas, I found myself ruminating over that visit to the library. Despite Karen's gracious invitation, I knew that I wouldn't return to explore the collection and the serendipity that it might offer. And yet I wasn't being deceptive, or even merely polite, when I told her that I, too, appreciate serendipity. I was being reticent. As a scholar, I have spent years trying to understand how organizing systems, like libraries, produce experiences of serendipity, curiosity, wonder, inspiration, and other forms of value beyond finding what you were initially looking for. I have taken Karen's role many times, earnestly trying to explain how organizing systems tell us stories as we use them—sometimes marvelous, unexpected stories.[1] That's the magic of serendipity: unexpected data stories.

And yet most organizing systems don't seem marvelous: they seem ordinary and dull. They don't seem to be telling stories. This is a contradiction that I relish, and so, if I were to start talking about it, I might, like Karen, rise excitedly from my seat as I explain that, when you encounter a book about the history of Denmark on a library shelf, you are coming into contact with a surprisingly intricate web of human judgments and decisions. For instance, it's a decision that libraries collect books, that books should be described by their subject matter, that history is a subject, that history is arranged according to nations, that the thing you are looking at is, in fact, a book, that this book you are looking at belongs in a library, and that it is a book about Danish history. Everyday ordering systems like libraries are so ubiquitous and conventional that such judgments seem obvious and unsurprising, not really decisions at all. We expect to find Danish history books on library shelves next to other Danish history books, and so we don't think of this as a decision that is telling us something vital—telling us a story—about history and libraries and books.

Serendipity happens when you find yourself, by choice or accident, listening to the story that an organizing system is telling. Sometimes, serendipity arises because decisions about selection, description, and arrangement are made exactly as you expect them to be made. The story, in other words, is in a genre that you recognize. You are looking for books about Danish history and you find, nearby, a book about relations between Denmark and Sweden; you realize that you're actually interested in Scandinavian history, and not just Denmark. But serendipity can also occur because organizing systems don't work at all as you expect: when the genre is strange and unfamiliar. Maybe, for instance, history books are arranged by the type of events they chronicle (political events, economic events, cultural events) rather than where or when the events took place. When you read this story, you might realize that you're not interested in a place, like Denmark, but in an idea, like collective action.

By this point, if I were having this conversation with you, I would probably be gesturing with my hands a lot, getting a little breathless with excitement. The library is just a convenient example! I would exclaim. What I'm talking about is everywhere! Any set of any documents, from government archives to hospital health records to your favorite music streaming service! Any dataset at all! Any store or marketplace, physical or digital! All collections have stories.

An Enchantingly Strange Museum

I might then look off into the distance for a moment before telling you about the Museum of Natural and Artificial Ephemerata, an odd and somewhat eerie collection inside an otherwise unremarkable family house in Austin, Texas.[2] The husband and wife curators who manage the Museum of Natural and Artificial Ephemerata have assembled a prodigious set of objects that they have arranged into exhibits such as Celebrity, Animals, Travel, and Sleep. Some of the items are perfectly normal (a tourist figurine of the Infant of Prague, like the one my sister and I bought as a souvenir for our Catholic mother many years ago), others are fantastical (a preserved flamingo head), and others are both at once (a vial of sleepysand collected from the curators' own eyes over a certain period of time). When you visit, the curators will guide you through each display and its wonders and, if you are lucky, you may see their cat jump through a hoop on command.

The MNAE is confounding. It subverts conventions for what a museum, as a kind of organizing system, is supposed to do. We commonly think of museums as preserving objects of value because they have artistic merit, or because they are historically significant, or because they have some educational purpose. The museum describes and arranges these objects in order to communicate their value to us (to show that a painting is in the style of German Expressionism or that a bronze pot is representative of the Shang dynasty). The items in the MNAE—a cigarette butt purportedly smoked by Marilyn Monroe just before her death; a taxidermy rabbit with antlers, identified as a jackalope; a small, nondescript puck of mud and animal hair described as a yeti toy; and a souvenir of a rabbi's sojourn in India—are purposefully dubious in their provenance and technically worthless. But if the MNAE violates expectations for what a museum should collect, it conforms to expectations for how a museum describes and arranges its holdings. In the stories that the curators tell, the objects in their museum are amazing treasures that have been rescued from misunderstood obscurity. It seems like these stories must be jokes, purposeful falsehoods meant to be funny. But the curators' commitment to the museum is too deep for mere jokes. They do not seem wealthy, and yet they have built, at undoubtedly great expense, a special addition to their house to contain this "joke" collection. What establishes value if not commitment, a decision that value exists? A visit to the MNAE demonstrates how a decision to describe and arrange things in a certain way can make objects precious.

I might scratch the back of my head, then, searching for a more relatable example. After a moment, I will snap my fingers and ask you to consider something more quotidian: the supermarket. As a cook who enjoys discovering new ingredients, I can describe distinctive features of assorted grocers in every city where I have lived. The cheap prices on excellent produce at the 22nd and Irving Market in San Francisco; the tremendous kosher-for-Passover section at the otherwise unremarkable QFC in University Village in Seattle; the superlative bulk section at the Central Market in Austin; the huge bags of greens and extensive array of country ham at King's Red and White in Durham, North Carolina. This is a type of organizing system in which I have developed particular expertise, which means that I am sensitive to small distinctions between collections, and I can take advantage of these differences to facilitate serendipity. Because I know if the variety of tea or vinegar at any particular market is paltry or exceptional, or if the selection of jam emphasizes certain characteristics over others—perhaps the focus is on international brands, or organic options, or local producers—I can appreciate when it might be a good idea to take a second look at the honey, even if I'm not shopping for any. Chestnut honey from Italy on my breakfast toast? If I do buy it, and come to understand its peculiar intensity and bitter edge, that will further hone my ability to recognize when serendipity has deposited a treasure in my path. This is how people read the structure of organizing systems and follow the trajectory of data stories.

And do you know why systems like Netflix and Amazon become so terribly frustrating? I might then bellow. It's because their rationale is commercial rather than curatorial, and because their inputs are statistical proxies for human assessments, rather than deliberated judgments. If Amazon and Netflix have such huge collections of resources, why do they annoyingly suggest the same selections over and over again? It's because they operate on old, tired patterns, rather than setting new ones. Good gracious, Amazon, if I've just read a book by Author X, what's the point of recommending to me the entire oeuvre of Author X? All that tracking data you have, and your decision-making capabilities are less robust than my local bodega. It's scandalous!

About here, I will have become slightly embarrassed about how long I have spoken, and my voice will falter. It is likely that I won't have conveyed what I really want to say, but my frenetic attempt may convince you that I appreciate serendipity and that I've spent a long time thinking about how organizing systems produce it. And yet when Karen gestured toward a set of library shelves that she believed might be interesting for me, I was indifferent. As I sat in my office, trying to make progress on my book ideas, I tried to understand why I had felt this way. Yes, I had gone to the department library specifically to pick up some holds. Still, I could have responded to Karen's invitation by deciding to return another time. My ambivalence couldn't be explained merely because I was intent on another task.

A Call Number That I Can't Find

With my hands hovering over my laptop keyboard, fiddling with my ring instead of writing, I recalled how, a few days before, I had gone looking for a book in the Media, Cognition, and Communication (MCC) department's library on my campus. I couldn't find the book, even though I had written down its call number. MCC was a large department in the Faculty of Humanities, and it was organizationally divided into five sections, or subdepartments. In a surprise to me, the library's collection was also divided according to these departmental sections. The book that I was looking for was an anthropologist's discussion of bureaucracy. Was it in the Philosophy section? The Communication section? The Rhetoric section? The Film and Media section? After failing to find it in any of these, I gave up. I left, convinced that I had written down the number wrong. Nope; the book was in the Education section, the one area that I had neglected to check (see figure 1.1).

Figure 1.1
The Utopia of Rules, the book that I was looking for, in its place in the Education section of the library.

Although I was looking for a particular item and not systematically reading the shelves for what they might tell me, I had reasoned from this experience that the section-specific library system works against certain kinds of serendipity. This effect arises because the collection associated with each section (or department) is quite small, totaling just a few racks. But the classification system for organizing the shelves is a general one, a variation of the Universal Decimal Classification, meant for large libraries with broad holdings. This disjunction—small collection with a general classification scheme—makes transitions unpredictable for subject matter outside of the collection's core focus. The book I was looking for was tangential to all the sections I searched, so browsing around the identified call number didn't reveal anything close to the topic that interested me (an understanding of rules and their effects on social and political life). When I learned that the university's current collection development policy for its libraries was to restrict purchases to specific requests made by faculty members, this effect appeared even more likely to me. Without a coherent selection plan, this kind of fracturing in areas beyond a section's core domain is inevitable.

As I twisted my hair around my finger, contemplating my now-sleeping computer, I told myself that this analysis had been the source of my fidgety demeanor with Karen. The structure of small department libraries oriented around certain well-defined fields wasn't optimized for my current interests; I was not the core audience. Still, this explanation didn't feel satisfying. Yes, it might have been awkward to share these thoughts with Karen, but every organizing system tells stories that resonate with some people and not with others. If that was what was bothering me, I was certainly overreacting. As I turned my attention to the books that I had checked out, I put the trip to the library out of my mind, but I wasn't convinced that I had fully comprehended what had occurred.

Sure enough, on Monday morning, as I returned to my office, I found my thoughts revisiting the department library. Over the weekend it had become gray and rainy, and I shook the drops off my rainbow anorak as I entered our hallway. Observing this, my colleague Trine apologized for the Danish summer. "Oh, but this is lovely," I countered, "compared to the awful humidity that I'd be suffering in North Carolina." I chuckled to myself. It's so human to grumble about the weather, no matter where we are.

Sitting down at my desk, I recalled how I had also grumbled recently. At the lunch table with my new colleagues, the day before my trip to the department library, I had complained about not being able to find that book on bureaucracy. Gitte had said dryly, "Oh, you didn't understand the way the system works?" I thought she was making fun of my ineptitude. And she was, but then she said, "No one understands it. It's very frustrating." Then Jack mentioned that he never checked out books in person anymore: "I just have everything sent here, no matter where it is." Everyone agreed with me, I took this conversation to mean. The organizing system for department libraries was annoying, and interacting with it was pointless.

When I had puzzled over it in my office, thinking carefully about not being able to find the bureaucracy book, I had been able to consider the structure of the department library system in a general way, to identify how it might facilitate serendipity for some people but constrain it for others. However, my initial reaction to the department library system had been less deliberate. Because the system's structure had not aligned with my goals, I had dismissed it as fundamentally lacking in value. I had failed to recognize that it contained any worthwhile stories at all.

I had done what we all do when we grumble about the terrible weather: I had universalized my own experience without thinking about it. I had substituted a question about personal preferences—"Is the department library telling a story that I want to hear right now?"—for the more general question: "What stories does the department library tell and how does it tell them?" When two colleagues had agreed with me that the department library system of shelf ordering didn't align with their preferences, that had confirmed my impression that the department library was ineffective and outdated: that its data stories were not worth anyone's attention.

Danish Supermarkets

I had been doing exactly the same thing, I suddenly realized, with the grocery stores in Copenhagen. Every supermarket seemed so small, and yet the space was taken up with random things: tins of cod roe and liver pate; jars of pickled red cabbage and beets; refrigerated, mayonnaise-based salads; candles, socks, and potting soil. Or so I grumbled to myself. But of course, those things weren't random at all.

The problem wasn't poor selection and incoherent arrangement; it was my orientation. The Danish supermarkets weren't aligned for my American habits, just like the university library wasn't optimized for my current project. As a result, everything seemed haphazard and full of strange gaps.

Truth be told, it was frustrating and unpleasant, rather than invigorating, to be lost in the kind of environment in which I was used to being an expert. I'm embarrassed to admit, for instance, that it took me five visits to one establishment to realize that they did actually have roasted red peppers—they were in the aisle with olives, artichoke hearts, and pickled garlic, and I had been looking on the aisle with pickled cabbage and cucumber (see figure 1.2).

For both the supermarkets and the university library, I had made a classic mistake. I had focused on a question of personal preferences and situations—"Is this organizing system telling a story that I want to hear right now?"—instead of a more general question: "What stories does this organizing system tell and how does it tell them?" In short, I had asked "Do I like this organizing system?" rather than "How does this organizing system function?"

How had I done this? After all, if we think about these questions in a basic sense, we can usually distinguish them pretty easily. We all enjoy certain things, even as we recognize their artistic shortcomings. For instance, I devour detective novels for my pleasure reading, even as I acknowledge that the plots are often recycled, the characterizations formulaic, and the writing merely serviceable. Nonetheless, I find detective novels tremendously enjoyable! Similarly, we might be able to describe and appreciate the qualities that literary critics employ to designate a certain novel—let's

Figure 1.2
Two aisles of pickled things at the Meny supermarket on Vermlandsgade in Copenhagen. The roasted red peppers are in the aisle pictured on the right, not in the aisle pictured on the left.

say *Middlemarch* by George Eliot—as an example of classic literature. Even as we appreciate *Middlemarch*, however, we might not delight in it. In the abstract, then, it doesn't seem too hard to make distinctions between *enjoyment* and *understanding*. We make these distinctions all the time. Enjoyment is personal and idiosyncratic, often dynamic and hard to pin down. Understanding how something works and appreciating its qualities is more intersubjective and stable.

When we attempt to bring this more general process of interpretation to the domain of collections and their data stories, however, there are some complicating factors. In our everyday dealings with organizing systems, our operational mode of understanding tends to narrow onto instrumental concerns—what we are trying to do right then—which makes it harder for us to recognize the range of stories that a system might be able to tell. On one hand, it's perfectly reasonable, when we consider our interactions with information systems, to think about accomplishing goals or tasks and whether our progress with those tasks is smooth and efficient. But when we think only in those terms, we are primed to misidentify small incongruences or frustrations as universal problems that must be solved rather than as communication differences to be negotiated. We become focused on "Does it work for *me* right *now*?" rather than "How does it work?"

When I think about my trip to the department library, I see how I failed to keep these questions separate. And if I'm being honest, the moral of this story—the reason that I keep coming back to it in my mind—is the recognition that such clear-eyed transcendence is impossible, no matter one's will or level of expertise. In the supermarket, my inability to find the peppers—and the fact that I was looking for roasted red peppers and chickpeas in the first place, rather than pickled beets and leverpostej—very much affects the way that I am able to think about how the supermarket as organizing system works, and how I am able to recognize the stories that it tells. Even as I perceive that this is happening, my positionality continues to affect my comprehension of the system. In this case, I'm an outsider trying to finesse an unfamiliar environment, but no one is exempt from these effects. Members of the intended audience have their own experiences to manage in trying to comprehend how systems work. A Danish grocery maven will equally find those habits, preferences, and expertise to interact with any attempt to understand how a particular Copenhagen market works, in the more general sense. None of us can transcend the personal, local, immediate, and instrumental in our interactions with information systems, even though, in order to understand their value more generally, we should do our best to try.

For many everyday information systems, like the library and the supermarket, it may seem unnecessary to complicate our lives by thinking critically about the particular question "Does it work for me?" and asking how that personal experience relates to the more general question of "How does it work?" After all, the stakes seem low. What does it matter if I dismiss the department library system or Copenhagen supermarkets as generally worthless when I may not be seeing the complete picture? Practically, it might not matter at all, at least for my immediate needs. It won't help me find the bureaucracy book or the roasted red peppers. Indeed, even if I make a good-faith effort to understand and appreciate the department library, I may never make any unexpected, wonderful discoveries there. However, if I were to redesign the department library system on the basis of a skewed perception of how it works, I might cause unanticipated difficulties for students following focused degree programs, to whose activities the current system may be more closely aligned. No organizing system will serve everyone equally, which means that decisions about design and use need to be taken with care.

When data seems increasingly to imply computation, libraries and supermarkets seem distressingly analog, and it may be hard to see their import in my adventure here. But anything I say about the library and the supermarket also applies to Netflix (a collection of videos), Amazon (a collection of data about goods, or a catalog of goods), Twitter (a collection of short utterances), Facebook (a collection of user profiles), Google (a collection of Web pages), and so on. Anything I say about the library and the supermarket, equally, applies to datasets like the Armed Conflict Location and Event Data Project (ACLED) or the Johns Hopkins Coronavirus Research Center database.[3] A trip to the library; a trip through YouTube; a trip through the coronavirus database; the topographies of datasets, physical or digital, show similar contours, and our interactions with these systems follow similar trajectories.

Let's consider ACLED for a moment. ACLED collects data about armed conflicts around the world, between different kinds of actors, including organized states, nonstate groups, and civilians. ACLED documents many episodes of armed conflict that also appear in terrorism databases, such as the Global Terrorism Database (GTD). In ACLED, however, the notion of a terrorist event doesn't exist. ACLED's data describes armed conflicts involving different configurations of actors. This is not a mistake. The researchers who manage ACLED believe that *terrorism* is not a valid analytic category. ACLED is like the Danish supermarket with multiple aisles of pickled things, where the roasted red peppers (the *terrorist* events) take on a different character as they occupy a space next to the artichoke hearts and pickled garlic rather than a space by the pickled cucumber. And the GTD is like the typical American supermarket, in which all the pickled items get tossed together with less regard for their differentiating characteristics. ACLED reconfigures the notion of terrorism, whereas the GTD reifies it.

Or that's the idea, anyway. As I read this over, after time has passed and I've rewritten it many times, it sounds a little pat. Is it really so direct, to navigate from supermarket shelves to database schemas? How *do* we perceive the data stories that surround us?

Might *this* be the motivating question for my sabbatical project? Aha!

Two Serendipitous Books

If I can yet appeal to serendipity, the two books that I had gone to retrieve from the hold shelf at the Information Studies library can serve as guiding exemplars for such a task.

One of the books waiting for me that day was the novel *Vertigo*, by W. G. Sebald. Sebald's novels are, on the surface, extraordinarily dull. A traveler goes from place to place and traces the associations of his thoughts in response to what he observes. In one of *Vertigo*'s four sections, an unnamed narrator, who is and is not Sebald, journeys from Vienna to Venice to Verona. In Verona, he suddenly feels afraid, and he cuts short his trip. Then, seven years later, he retraces his steps. Although nothing much happens in a conventional sense, Sebald's narrator is constantly relating places to both personal and historical people and events, so that trivial moments of everyday life (losing a passport, eating a pizza) become mysteriously evocative of some larger significance. Everything is related to something else, and those relationships are transmitted through the material structures of related documents: receipts, passports, newspapers, frescoes in chapels, photographs.

The second book waiting for me was Walter Benjamin's unfinished *Arcades Project*. In this messy set of notes, sources, and commentary, Benjamin considers the Parisian arcade, a glass-roofed passage that enabled nineteenth-century pedestrians to stroll comfortably along connected storefronts, away from the carriage traffic on the actual street. The arcade is an organizing system that relates people and places and encourages certain kinds of interactions between them. For instance, to take advantage of the pedestrian activity that the arcade structures, shops begin to specialize and develop intricate displays. The displays, in turn, encourage different relations between people and goods, and between goods and social and cultural life. The arcade also functions, for Benjamin, as a conceptual motif that goes beyond what happens there; it stretches out to encapsulate a spirit or sense of the nineteenth century as a historical epoch, one that can be crystallized in the compressed and concrete figure of the arcade but cannot be easily expressed in words. Moreover, the text of *Arcades Project* is itself not a standard narrative but is instead a kind of arcade or passageway between distinct but related themes.

Both of these books use extended meditations on concrete, everyday details to explore how different kinds of structures—material structures, conceptual structures—work to generate meaning. Both of these books employ, as well as comment upon, the analytic and expressive powers of arrangement. In their own structures, both books are episodic; they leave a few threads hanging, for the reader to weave or not. But the loose strands of plot are kept anchored by an ever-present narrator whose persona both gathers and refracts the stories that they relate.

These books are touchstones because, in focusing so deeply on what might initially appear to be tedious minutiae, these books perform the double function of accessing the abstract through the concrete and of illustrating the multiple stories—from the boring to the wondrous, the frustrating to the fascinating—that even the most mundane travels—eating a mediocre pizza in Verona, strolling back and forth in a nineteenth-century shopping mall—can reveal. The very close narrator is given distance through the inclusion of quotations, photographs, and references, so that the relation between story and narrator is itself constantly in focus. The structural combination of common, everyday situations filtered through the specific experiences of a peculiar narrator prods the reader to reconsider what they might otherwise let pass by without introspection.

So, then, a book of everyday data adventures; a book that is also, in its heart, a book about reading and writing. To honor its origins—both literally, in a concrete personal experience, and figuratively, in an invocation to my scholarly forebears—it should begin with a trip to the library.

And so it has.

REFLECTION: HOW DO WE ATTEND TO DATA IN THE EVERYDAY?

The Intellectual Heritage of Knowledge Organization

In the field of information science, serendipity has been investigated from various perspectives. One approach looks at serendipity in the context of information-seeking behaviors and practices: how people think and act with information. Serendipity emerges here in discussion of browsing, encountering, and other behaviors in which search goals are less sharply defined.[4]

Another information science approach examines serendipity in the context of classificatory structures, such as bibliographic classification schemes, that facilitate activities like browsing and encountering. This perspective, drawn from the field known as *knowledge organization*, focuses on the design and assessment of classificatory structures themselves rather than the people using these tools. Serendipity enters knowledge organization through discussion of useful relations between classes, as in the library classificationist S. R. Rangathanan's (1957, 1959) motivating principle of helpful sequence, somewhat mystically elaborated upon as the alien, penumbral, umbral, penumbral, alien (APUPA) pattern. As a more recent example, Birger Hjørland's program of domain analysis asserts that schools of thought within an area of scholarship should form the basis of an organizing system, so that information seekers are made aware of how documents fit within scholarly conversations and debates (see, among others, Hjørland and Albrechtsen 1999; Hjørland 2002, 2004).

In knowledge organization, classificatory structures typically use *aboutness*, or the subject matter of documents, as the primary organizing principle. It has long been assumed in Western intellectual traditions that topical arrangements facilitate both directed retrieval and serendipitous discovery. From the laborious compilations of early modern bibliographers, to the comprehensive systematization of early twentieth-century library classificationists, to the specialized technical vocabularies of post–World War II indexers, classificatory structures for information management have centered on aboutness.[5] Nonetheless, the concepts brought to bear throughout the long history of subject classification are easily transferable to other areas—like supermarkets and Netflix—that are not documents in the traditional sense and to which aboutness does not directly pertain.

More generally, it is my conviction that the scholarly traditions of information science and its antecedents, such as documentation and librarianship, can usefully inform upon matters that we now perceive as data related. Older literature may seem narrow and old fashioned in discussing documents rather than data or in setting out rules for human describers rather than rules for algorithmic datafication. Nonetheless, if we disengage the fundamental concerns of these older analyses from their unfamiliar sociotechnical arrangements, they continue to be relevant and perceptive. For instance, in 1968, the philosophically trained information scientist Patrick Wilson argued convincingly that there is no sound logical principle upon which to establish and relate the constituent objects of the bibliographical universe. Any means that we might choose to define what documents are and how to relate them, Wilson argued, is provisional and conventional. In other words, there is no consistent, universal basis upon which one can make decisions like "the subject of this book is history, and not political science" or "the Spanish translation of this book is, effectively, an entirely different book." Wilson's arguments remain just as trenchant when we shift the vocabulary from that of books and documents to that of information and data. Using more currently familiar language, we might summarize Wilson's exegesis as "all data is some person's story about the world" or "all data is inherently biased."

As another example, in 1976, the information scientist Marcia Bates wondered why bibliographies did not inform users of the specifications that librarians had relied upon in determining their rules of selection, description, and arrangement. Here, Bates extended Wilson's 1968 arguments to observe that, although data creators often assume that the rationale behind their decisions is apparent and interpretable to others, data users invariably lack the context to understand data as the creators expect and intend. Bates noted incisively that "it is not enough to say that a bibliography is on trees if it in fact has been defined to include shrubs, or if it is meant to cover only material on tree species and not to cover ecology of trees" (Bates 1976, 19). In today's vocabulary, we might express Bates's proposal like this: if all data is inherently biased, then people who create data should attempt to make their decisions and decision-making processes transparent for the users of that data. In other words, they should lay bare the authorship of their data stories. When I talk about supermarkets and Netflix, my ideas still rely upon the traditions of information science, as exemplified by Wilson and Bates.

Reciprocal Themes across Multiple Disciplines

More recently, scholars of digital humanities, science and technology studies (STS), and critical data studies have made similar arguments. There is widespread agreement among these scholars regarding the nature of data as always already "cooked" rather than "raw," the impossibility of unbiased algorithms given the impossibility of unbiased data, and so on.[6] From my perspective, researchers from these fields tend to be less interested in the details of data cooking: the fundamental techniques by which we establish what *things* are and how to describe them. This is where the scholarship of information science retains its distinctive purchase.[7]

Scholars outside of information science are often introduced to classificatory matters via the groundbreaking *Sorting Things Out* by sociologists Geoffrey Bowker and Susan Leigh Star (1999). Bowker and Star are interested in classification schemes as representative examples of sociotechnical *infrastructure*. Using cases such as the International Classification of Diseases (ICD) and race categories in South Africa's apartheid system, Bowker and Star emphasize the role of classificatory structures in shaping social practice. The emerging field of infrastructure studies owes much to Bowker and Star.[8] Infrastructure research might examine physical infrastructures, like undersea data cables, or intangible infrastructures, like metadata standards for scientific observations. Like Bowker and Star, infrastructure scholars often direct their attention toward the social worlds in which infrastructures are enmeshed rather than the particulars of infrastructure design. For instance, the ethnographer David Ribes (2019) is interested in the concept of a *domain* as a mechanism through which disciplinary scientists (such as geologists) negotiate collaborative work with partners from computer and information science. For Ribes, a *domain* is a social construct through which one can understand how classificatory infrastructures come to be. This perspective on domains is subtly different from the work of knowledge organization researchers like Hjørland, for whom a *domain* provides a rationale by which classificatory infrastructures are designed, implemented, and assessed.

More broadly, data and the stories that it is and is not able to tell have increasingly become objects of general concern. Following the 2016 elections in the United States, revelations regarding the use of social media profiles to disseminate targeted misinformation masked as political advertisements catalyzed public interest in organizing systems that focus attention on *this*, as opposed to *that*, to unsubstantiated rumor and innuendo (so-called fake news) as opposed to verified reports based on reliably sourced evidence. Subsequently, the role of data in driving algorithmic decisions to surface certain content and bury other content, or to promote certain outcomes over others, has come under increased scrutiny. For instance, journalistic and scholarly investigations have identified algorithmic processes that cause harm to vulnerable populations by relying on data that reifies existing social stereotypes.[9] Increasingly, research seeks to describe how data works to depict reality in certain ways as opposed to others.[10] That such studies often focus on the impact of certain data stories upon marginalized groups demonstrates the pervasive and continued use of data as an instrument of subjugation, both purposefully and unintentionally.[11]

Comparatively little research examines the value in localized, particular data structures. My own inspirations have often come from poetic, rather than analytic, sources.[12] This may be changing, though: a burst of recent scholarship such as Loukissas (2019), D'Ignazio and Klein (2020), and Costanza-Chock (2020) emphasizes an approach to data analysis and design that foregrounds situated, local perspectives. Notably, these books encourage their readers to reimagine and reinvent the systems they encounter.[13]

Information Design, Power, and Accountability

I can trace my own interest in the serendipitous potential of information systems to a seminar that I took as a master's student at the University of California, Berkeley, with Professor Nancy Van House. In Nancy's seminar, we discussed an article by anthropologist Lucy Suchman in which Suchman struggles to comprehend her role as an ethnographer at Xerox in the 1990s (Suchman 2002). Suchman's technologist colleagues expect that her research to understand specific communities of practice will inform design work across contexts. Suchman attempts to reconcile her own understanding of what she is doing—ethnographic description of particular work environments—with what her colleagues want—implications of those observations for design. For Suchman, distilling such design implications seems an impossible project, because how can a partial understanding of one context lead to some kind of generalizable rule? But what is an anthropologist at a technology company doing if she is not helping to inform the design of products? Ultimately, Suchman suggests that a feminist-inspired notion of partial perspective—for which

Suchman draws on Donna Haraway's (1988) essay on situated knowledges—does not align easily with a culture of technology production focused on the design of universal solutions.

In our seminar, Nancy pointed us to a passage in which Suchman describes her own self-doubt. Was it, Suchman wondered, a "personal shortcoming" that she could not provide the general design guidelines that her colleagues asked for? At the time I wondered why Nancy kept emphasizing that passage. But fifteen years later I return to that passage as well. People are constantly asking me to provide them with rules for ensuring good data. These requests make me very nervous. *Good data* is an elusive, amorphous notion. But this is not a satisfying answer, and if I tell people this, I risk being bypassed as useless or dismissed as incompetent.

When I read Suchman now, I realize that when she is asked for design implications, she is also being offered a kind of power. Refusing that power is both very hard and, in our society, very weird. What is supposed to happen then? Suchman cannot say. If you are really trying to figure out what it means to act according to feminist values, you often cannot say, which remains an almost unaccountable state of affairs. And refusing power seems irrational. Logically, I can see many arguments, both idealistic and cynical, that justify providing the design implications anyway. There is, also, the emotional aspect of being treated like a nonentity. I am particularly susceptible to that.

None of this is new. Daniela Rosner's *Critical Fabulations* chronicles a similar kind of struggle to reconcile feminist values with design practice, and there are many others (Rosner 2018). But the question of what to do remains vexing. If I refuse to prop up existing arrangements by providing "best practices," am I making things better or worse, and for whom?

At my heart, I am an idealist, and idealists often have a complicated relationship to power. The disciplines of librarianship and information science have long aligned themselves with what we now call social justice. Librarians of past generations wanted to make society better, just like today's librarians do. Today, however, we would find the goals and methods of past librarians to be, at best, misguided. Nineteenth-century librarians confidently believed that the working classes could be lifted up, morally and intellectually, through access to the *best* books (Wiegand 2015). Determining what the best books might be, and what a moral person should be, was not a matter for debate.

More flamboyantly, the Belgian classificationist Paul Otlet believed that world peace could be achieved through universal access to *facts* (Rayward

1994). For optimal efficiency, facts would be extracted from the cumbersome books that contained them and made available through a meticulously worked-out classification system of precise specificity. The possibility that one might disagree on what facts are or on how to classify them was not something Otlet contemplated. If we think of Otlet's project as one person telling everyone else what *the facts* are, then it seems horrifying. But, as Boyd Rayward (1994) emphasizes, Otlet was motivated by idealism. He was after nothing less than world peace, and he devoted his life and fortune to the cause. How can we not admire and respect Otlet's singular vision, commitment, and tenacity? At the same time, Otlet's position on facts could easily be a tool of oppression, rather than liberation.

If Otlet's grand aims have resonance for us today, the instruments by which Otlet would create such change—bibliographic classification systems! cabinets of bibliographic references arranged on standard-sized paper slips!—seem incredibly humdrum, ill-aligned to such disruptive goals. Along these lines, Jutta Haider and Olof Sundin (2019) insightfully remark upon the strange paradox inherent in our daily discourse regarding information systems (specifically, search systems): information systems are seen as vitally important and integral to a democratic society, and yet information systems are also considered to be so commonplace and mundane that no one needs to put much effort into using them critically and well. What results is a weird and powerless situation in which there is a lot of talk about how much influence information systems have over our daily lives, and yet our collective ability to identify and respond to this situation remains weak—and, crucially, our will to address this situation likewise remains weak. We simply do not believe that using Google (or a library, a supermarket, or database of conflict events) is an activity that requires skill, knowledge, and technique—any sort of understanding of *how it works*.[14] In Julian Warner's (2009) Marxist model of information retrieval, this dissonance results from a failure to appreciate the human mental labor required by users of search engines, upon whose brains all the semantic labor of evaluating and assessing vast lists of results rests.[15] Analyses such as Warner's, and Haider and Sundin's, suggest that it is insufficient to say that attention must be paid to the workings of information systems; *techniques* for paying attention are required. The essays in this book attempt to demonstrate some of these techniques.

2
OBJECTIVITY

This chapter focuses on quantitative measurement. Because quantitative data is conventionally associated with objectivity, and because objectivity suggests validity, we tend to trust numerical measurements, even as we are generally aware of their uncertainty and instability. I scrutinize data conventions associated with objectivity and consider the effects of disrupting those conventions.

In the adventure essay, I tell stories about

- Stepping on a scale.
- Step counters on smartphones.
- The allure of numbers.
- Measurement and convention.

In the reflection essay, I ruminate over

- The two cultures of the sciences and the humanities.
- Human judgment in quantitative measurement.
- The conduit metaphor and the mathematical theory of communication.
- Objectivity as that which speaks for itself.
- The human work of data collection.

ADVENTURE: THAT SCALE MUST BE WRONG!

Stepping on a Scale

In 2018, my partner Jason presented me with a very thoughtful Christmas gift that was also rather embarrassing. He proposed to send me on a spring break trip to visit my sister in Manhattan, so that I might, with all of New York to search in, successfully buy a new pair of pants.

Throughout the autumn of that year, Jason had been listening to me complain about how tight all of my pants were becoming and how I didn't want to wear them anymore. How had this happened? I constantly whined. I've been attacking the climbs in spin class like a beast. I've been lifting more on my squats and bench presses than ever before. And yet whenever I weighed myself on the scale at the gym, the number surprised and disappointed me. Maybe the scale wasn't calibrated correctly, I would tell myself. When I got similar results on different scales at different branches of the gym, maybe I was wearing a particularly heavy T-shirt that day. Maybe I had drunk an excessive amount of water. Maybe I had eaten a little too much cake that week, but that was unusual. Really, I don't normally eat that much cake . . . hardly any cake at all. Maybe the gym had purposefully adjusted the zero point on all its scales to encourage its patrons to visit regularly. Ah, that was probably it.

Clearly, my incessant grousing had started to drive poor Jason insane. Chastened, I agreed to his plan. I went to New York for spring break. I bought some new pants.

Then, in June 2019, we moved to Denmark for my sabbatical. It took a while to settle in, and when I wasn't at the university, Jason and I were out running errands or just seeing the city. It was two months before we joined a local gym and finally went to work out for the first time. Argh, I was so weak! My lifting capacity was pitiful, and I had little stamina on the cardio machines. As I dragged myself into the locker room, I noticed a scale in the corner by the showers. Against my better judgment, I stepped on it.

I squinted at the display, and then I double-checked the conversion from kilograms to pounds on my phone. Surely that measurement was too low? Maybe I hadn't drunk enough water that day. Actually, I hadn't been drinking enough water in general, so that was clearly a factor. I was dehydrated. And my new gym shoes were definitely lighter than my old ones. No, it was probably the scale, which seemed cheap and flimsy. All the equipment at the gym was rather old, in fact. Yes, that scale must be wrong!

Good grief. Why did I keep doubting these measurements, whether they went up or down? Why was it so difficult to accept the numbers right in front of my face? Surely, all the scales in all the gyms of the world couldn't be wrong.

Or could they?

When we talk about the accuracy of particular measurements, what are we really talking about?

Certainly, it's not irrational to assume that our everyday measuring devices are likely to produce tolerably correct results.[1] When I use the scale at any gym, the numbers that appear are clearly not random: they have a correspondence to what I might expect from the object being placed on them (me). No scale has ever reported my weight as 500 pounds or as 5 pounds. Furthermore, each scale is consistent in its operation. At no time have I stepped on a scale twice in succession and seen my weight change from 120 pounds to 220 pounds. It is reasonable to believe that the scales that I typically encounter generally function according to their established level of exactitude. On the other hand, in everyday life, we seldom investigate the precision of our measurement devices, particularly when they seem to have authoritative digital readouts.

Step Counters on Smartphones

Take, for instance, step counters on smartphones. Try going for a walk with a few friends and then checking everyone's phones to see the distance traveled. You're likely to see variation between the measurements.

It's easy to think of reasons why step counters might display minor discrepancies in their totals. Some deviations arise from differences in the phenomena that are recorded; that is, from differences in how people walk, which affect the absolute distance traveled. A shorter person generally has a shorter stride and takes more steps than a taller person. Some people move from side to side as they walk, which results in more steps taken and more total distance.

Other properties of walking, such as speed, cadence, and smoothness, can affect how the phone processes the data that it receives. In response to a particular cadence, the phone's precision and sensitivity may diminish. For instance, a phone may calculate distance differently for a person who walks with an even, slow step from the way it does for someone with a quick, bouncy step. As a result, the same phone might record different values for the same distance traveled by different people. Differences between phones add complexity to the equation. Hardware and software components may respond to variation in these properties to a greater or lesser degree, so that a Samsung Galaxy phone with the Android operating system may process a bouncy step differently from an Apple iPhone with the iOS operating system. Many such factors might affect how a phone collects and interprets step data: whether the phone is carried in a pocket, in a backpack, or in one's hand, for instance. If we think about it for a few minutes, we can come up with a sizable list of circumstances that might play a role in different counts.

But not all variation can be easily explained. I had two iPhones during my time in Copenhagen: a work phone and a personal phone. After walking what seemed like a lot one day, I wanted to see the distance I had traveled. Curious, I checked each phone (see figure 2.1). The distance was not the same, and yet it was the same type of equipment with the same software, carried by the same person in the same receptacle.

I am perfectly aware that step counters are approximate, and I knew before I looked that the numbers on each phone would probably be different. Yet I still found it startling and almost eerie to see the differences side by side. And, truth be told, I was reluctant to change my behavior toward these numbers. Was it now impossible to look at my phone and feel confident that I had walked 10.6 miles that day, or to congratulate myself for walking more than ten thousand steps every day for a week? Of course, it had always been impossible to take those numbers at face value, and yet I had cheerfully done it anyway. Indeed, even as I wanted to exploit the

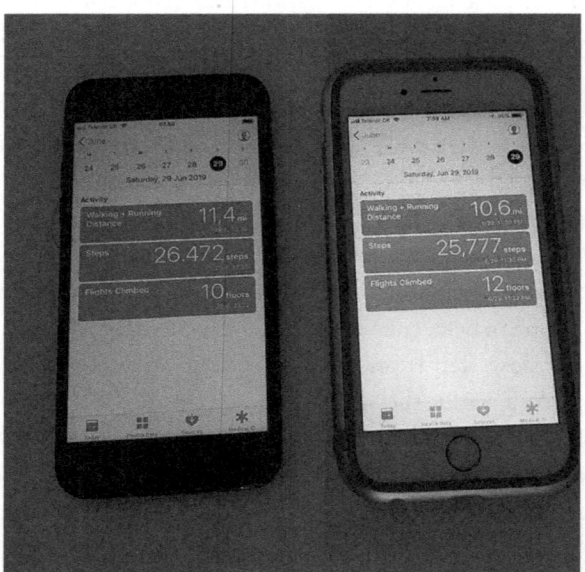

Figure 2.1
The same walk captured by two different iPhones.

uncertainty of measurement with the scale in the gym, the prospect of similar uncertainty in the step tracker flooded me with ambivalence. I wanted to be able to rely on the step tracker's numbers, because those numbers told stories I was happy to hear, such as "With all those steps, I can definitely have as much cake as I want at all times."

With the scale, as opposed to the step tracker, I was compelled to develop arguments that would destabilize the utility of the measurement. My motivation may have been fueled by self-interest; I didn't want to be gaining weight, because I do like eating cake, and I didn't want to choose between adjusting my cake-eating habits and buying new pants. But the arguments that I formulated to test and interrogate the potential uncertainty of the scale's data are widely applicable to many situations and not at all irrational. Some of my arguments questioned the device performing the measurement (was it cheap or old or accidentally calibrated incorrectly?). But most of them pointed to other factors involved in the process of creating data values. What was actually being measured (how did clothes and shoes or the lack of them figure into the total)? How did the measurement process affect the results (had any scale been calibrated a certain way on purpose, and was I converting between measurement scales accurately)? Other arguments pondered an even more fundamental issue: what did the measurement represent, anyway?

In everyday life, we tend to treat measurements like weight as being expressive of some regular state: what you weigh in a general sense. Most of us, when asked to report our weight, are likely to offer a representative number, or perhaps an ideal number—exactly how we interpret weight as this kind of regular state can vary. What we are less likely to do, when asked What do you weigh? is to report exactly what we last saw on a scale. We are unlikely to do this, because we know that the number will probably not be the same when we weigh ourselves again (whether or not we are actively trying to gain or reduce). Weight is not a fixed property, and so from a certain perspective any individual measurement of weight, no matter how exact, is approximate. It's approximate because it is only accurate for a certain moment in time, and that moment might be more or less similar to other moments in time, for any span. Does it sound a little hysterical to think about how a weight measurement might change based on one's water intake? It shouldn't. We should all be wondering what any particular measurement actually means in relation to the ongoing existence of that object.

Usually, however, we only expose the uncertainty around measurement when we experience some kind of conceptual breakdown, that is, when something seems wrong: I can't weigh that much!

The Allure of Numbers

When a measurement is revealed as inherently uncertain in its relationship to the world as it appears to be, a common response is to take more measurements. If any one measurement is too reflective of particular circumstances, then we might weigh ourselves every week, or every day, or three times a day, or constantly throughout the day. If there is variation with each measurement, then the trends, or the average, or some means of statistically comprehending our enormous set of measurements will be more . . . well, here is where our notion of accuracy gets complicated. Each measurement might be completely accurate in terms of the precision with which it captures the property that interests us. It's just that the relationship between these individually accurate measurements and the representative (or is it ideal?) state of the object being measured is rather ambiguous. Still, if our concern is that a single data point is not fully expressive of the generally persistent state associated with a dynamic property, then collecting more data does appear to be a valid strategy. But what are we doing when we collect more data? Does more data solve the problem of uncertainty around any particular measurement value?

Let's think about it. Imagine that I have sensors in the desk chair where I work, in the sofa where I relax, and in the bed where I sleep; those sensors constantly weigh me, gathering a tremendous dataset. We can do a lot of things with this data. We can generate simple descriptive statistics to express the average value, such as the mean, median, mode, and range. We can look at trends and patterns, to establish how the data varies over time and to predict how it might vary in the future.

All this can help us to understand the contours of uncertainty in the data—the extent and pattern of fluctuation versus stabilization—but none of it banishes uncertainty entirely. Let's take the notion of an arithmetic mean, for instance. The mean, as a value, may never have been actually recorded—it is, in some sense, a product of the imagination. If I sometimes weigh 75 kg and sometimes weigh 70 kg but never (or very seldom) actually weigh 72.5 kg, is it adequately expressive of reality to report my average weight as representative of my general state, even if this number is, in a real sense, completely inaccurate? Similarly, the median value may be rare; the mode, extreme. All these concepts together indeed provide a more coherent picture, but it's a picture that accommodates uncertainty rather than excluding it. And although the past is often an excellent predictor of the present and the future—sometimes it's not. I moved to Denmark for a year, and the established pattern was disrupted. If I hadn't moved to Denmark, would that have happened? Who knows? Indeed.

But if my reaction to stepping on the scale was in some ways eminently sensible, in other ways it was remarkably deluded. What did I hope to learn from this information? I already knew that my pants were tight; that wouldn't change if the number that I observed on a scale was higher or lower. I already knew that my muscles were getting stronger due to my sweaty exertions in spin class. Although there was probably some kind of association between weight and the concepts that I really cared about—the fit of my pants, the capacity of my strength—the relationship was clearly not a direct one. Why keep stepping on the scale at all?

Well. I am an unabashed humanist, and even I find numbers to be beautifully concrete and definite.[2] For most of us, numbers provide a feeling of surety. If someone asks us how many apples are in the fruit bowl, we can count the apples and feel that we have definitively answered that question. We might argue incessantly about what particular numbers mean, for sure. But the numbers themselves are clear and easy to grasp. Let's say there are six apples in the fruit bowl. Is that enough apples to make a pie? We might disagree about that (it would depend on the size of the apples and of the pie dish and on your preferences for how much filling a pie should have). If you eat six apples, have you eaten a lot? We might debate that (does *a lot* refer to the volume of the apples or to their energy content?). Is it the same, nutritionally speaking, to eat six apples as to drink the juice of six apples? We might have different views. But the idea of the number *six* and how it applies to apples? That we are unlikely to question or contest. This kind of agreement gives us the power to make firm decisions.

Numbers are, moreover, hard to deny. If my pants feel a little tighter one day, I can dismiss this perception as imaginary—all in my head. But although I can argue with a number on a scale for all the reasons that I mentioned earlier, I can't pretend that it didn't exist. Whatever it might mean, the number was there. Numbers, as well, are easy to communicate. If I complain to my partner Jason that my weight on the scale at the gym has been increasing over the last month, then he has to acknowledge that something is actually happening, even if he thinks my reaction is absurd. The communicability of numbers further manifests as the quality of being actionable. Numbers can be easily associated with goals or standards so that we understand the goals consistently. As a directive, "Walk more" can be interpreted in different ways, but "Walk ten thousand steps a day" seemingly cannot.[3]

These captivating qualities apply to any discrete measurement scale. In the university, for instance, we deal a lot with grades. In the United States, grades are often associated with a letter scale rather than a numerical one, but the allure is similar. Students, especially, find grades mesmerizing. For instance, at the University of North Carolina at Chapel Hill, where I currently teach, graduate students do not receive traditional letter grades as the final mark in a course. Instead, graduate students receive a P (for pass), an H (for high pass), an L (for low pass), or an F (for fail). In practice, H grades are rare (maybe 5 percent of the class), and L and F grades are exceedingly rare. Almost all students receive a P. There is no concept of a grade point average or of any similar quantitative comparison of relative student performance. Consequently, the actual number of points that a student might receive for any particular assignment within a course is essentially meaningless. Did you receive 90 out of 100 points on your taxonomy project? Did you receive 70 out of 100 points on your taxonomy project? It doesn't matter; at the end of the course, you are getting a P! Nonetheless, students react more strongly to meaningless point totals than to carefully prepared written comments. "I only got 91 out of 100," I might hear a student say, dejectedly, as they ignore all the praise that I have written about their project.[4]

Measurement and Convention

Nonetheless, even as I find grades distracting and troublesome, I have continued to use them. Indeed, I have developed systematic grading criteria for my course projects, assigning points to each criterion. My whole assessment structure, which I am theoretically free to develop myself, is based on the assignment of points (see figure 2.2). Why do I do that?

I assign points because everyone expects them: students, other faculty, and administrators. Our course management system and other technical infrastructure expect points. The form of grading rubric that I employ, in

Objectivity

Descriptive Schema

Student name:

Grading criteria	Possible points	Points earned
The following are clearly described: what constitutes a member of the defined set of entities, the schema's audience and purpose, and how an entity should be identified and distinguished from other entities.	17	
The defined attributes effectively represent the selected entities in the context of the described purpose, and the value space effectively represents the extent of the attributes.	17	
The documentation is sufficient to describe actual entities accurately and comprehensively within the context of the selected purpose.	17	
The critical reflection thoughtfully considers the design process, product, or both, using the experience of creating the descriptive schema to productively engage larger issues of theory and practice (that is, the reflection does not merely summarize or justify the design process or product; it interrogates it).	17	
All project components follow a logical document structure, are clearly written, and use correct grammar and punctuation.	16	
All the project components are included.	16	

Wonderful things about this schema:
-

Things to improve for this schema:
-

Total points: /100

Figure 2.2
An example of one of my grading rubrics, with criteria matched to possible points.

which points are mapped to grading criteria, was introduced to me by pedagogical experts at my former university's teaching resource center. And, to be honest, even as I grumble about grades, I, too, expect them. Grades seem like a commonsense part of everyday life, like apportioning one's food intake into three meals per day. In short, grades are conventional, much like breakfast, lunch, and dinner. Social conventions like grades and mealtimes might be arbitrary and unnecessary, but they can be difficult to ignore. On one hand, there's nothing stopping you from eating at variable times throughout the day, whenever hunger strikes. On the other hand, it can be socially disruptive to adopt this behavior. Your work might schedule meetings around a certain lunch break; your family might eat dinner together at a certain time; restaurants might be open during certain hours. Grades are similar.

The quantitative aspect of grades perpetuates their conventionality. I suppose that what I really mean is that the expression of grades as a form of quantitative measurement is conventional. It is, for instance, because grades are associated with quantitative scales that we believe that we can use them across multiple systems (multiple courses, multiple departments, and multiple universities) and trust that the values are relatively equivalent. Just as six apples in one fruit bowl are the same as six apples in another fruit bowl, a 90 on one exam is the same as a 90 on another exam, and an A in one course is the same as an A in another course. And just as with the apples, this is perfectly true at the level of numbers and balderdash at the level of actual meaning. The six apples in one fruit bowl are Macintosh apples and not at all good for making pies; the six apples in another fruit bowl are Winesaps and should bake up well. The A in one course required many hours of intense work; the A in another course required little effort. Six is six and A is A, but six apples are not necessarily six apples, and an A in one course is not necessarily the same as an A in another.

Nonetheless, we often choose to concentrate on quantitative similarity—six is six—rather than qualitative difference—Macintosh apples have a different taste and texture from Winesap. The quantitative seems more solid and reliable—hard to argue with, even if limited in meaning. My grading rubrics illustrate this. Students trust the quantification that the rubric implies, even as the numbers provide little insight into their performance. The comments that I write, on the other hand, are less resounding, even less trustworthy. Before I used the grading rubric, students would occasionally argue for more points. Why 90 and not 95? they might have asked. With the rubric's additional quantitative overlay, students never question the points that they receive. Is my process different with the rubric? No. But when I use the rubric, I am making it seem as if assigning grades is more of a mechanical process (like stepping on a scale) and less of a human process (like contemplating the fit of one's pants). This mechanical aura makes the rubric and the grade associated with it seem more objective than written comments. It is this aura of objectivity that encourages us to forget what we know perfectly well from everyday life and to pretend that six apples are six apples, 90 is 90, and A is A. Now, in practice, assigning grades is clearly not like stepping on a scale, and the rubric doesn't change this. I'm doing just as much human interpretation when I assign five numbers for each grading criterion as when I write comments. But the rubric's aura of objectivity decreases the propensity of students to argue with me.

I adopted my grading rubric to increase transparency and fairness, not as a rhetorical trick to quell dissent. Nonetheless, the rubric exploits and perpetuates social preconceptions about what grades and measurements and trustworthy data in general are supposed to be: quantitative and objective. Moreover, I benefit from this arrangement. The rubric makes the process of grading easier for me. In my defense, I could argue that there is little harm in my use of the rubric. My grades are the same, and the points are meaningless anyway, after all.

However, it's through just these sorts of little decisions that specific measurement schemes, such as grades for courses and weights for people, and more general descriptive tendencies—quantitative measurement and putative objectivity—maintain social power. My use of the grading rubric contributes to a conventional, everyday understanding of what good data should be, in which we focus on the certain aspects of numbers (the ways in which six is six and A is A) and diminish the uncertain aspects of numbers (the ways in which six apples are not the same as six other apples, and an A in one course is not the same as an A in another). If we are troubled by the rampant application of this perspective, for instance, to algorithmically decide jail sentences or bail amounts, then we have to acknowledge that such developments do not arise just because the technology is available. They arise because the prevalent cultural conditions enable us to see and to manipulate data in certain ways. If my grading rubric legitimizes this culture of quantification, it is not harmless.

I could, of course, just stop assigning grades to my class projects, and provide only comments. But the costs would be high. Assigning points to student work is easy. Writing comments is time consuming. Moreover, such work is not highly valued. If I produced fewer scholarly publications in order to avoid assigning grades, I would be castigated for mismanaging my time. And honestly? I don't enjoy producing that kind of detailed feedback. It's a slog.

These may sound like petty complaints, but they are not. Everything that is made has a cost, and data is no exception. Sometimes we find great value in laboriously created, handmade items. But would it be worth it to manufacture bespoke paper clips or light bulbs or toilet paper? We would probably find such efforts misplaced. The low cost of quantitative measurement is a tremendous advantage. Assigning points to student work is easy and cheap. Stepping on a scale, too, is easy and cheap compared with soliciting (and hearing) human opinions about the figure you cut in those somewhat tighter-fitting pants.

So we need to decide whether the harm caused by grades and scales is worse than the allure and advantages of quantitative measurement. The harm is in the distorted view of reality that quantitative measurement propagates: a focus on simple numerical comparison rather than the circumstances in which the numbers gain meaning. The advantages of quantitative measurement are legion: simple to understand, conventionally accepted, easy to generate, cheap to produce.

The challenge of this calculation is that the harm caused by an individual manifestation of *grading* or *stepping on a scale* is vastly different from that caused by the infrastructural systems in which grades and scales are embedded. If I look only at my own courses, the costs of some alternate data enterprise seem unwarranted. If I look only at my own life, the harm of annoying my partner Jason with my talk of scales and rising numbers is small. But the harm of a grading culture, and the harm of a body-weight culture, with all the accompanying sociotechnical systems in which those norms are enforced, is much greater.[5]

Changing a culture is a major undertaking, and a data culture like grades or body weights is no exception. But thinking of these issues as cultural in the first place can help to open the imagination.[6] When I teach information organization to master's students, the first project that I set them is to design a descriptive schema: a set of specifications for generating data about some group of things. They can choose to describe whatever they want—coffee beans, computer programming languages, or mythological beasts; it doesn't matter. Initially, everyone thinks this project is beneath their capabilities: a rote task. When I explain that the whole point is to treat the description of landscape paintings or laptop computers as an open design problem rather than a reification of convention, my students are dubious. How else would we describe science fiction movies if not in the way that Netflix describes them? How else would we describe pain medication if not in the way that pharmacies describe it? The students think that data is a matter of describing things *as they are*, and that there is no art to it and certainly no fashion. They don't think that data is the same as a pair of pants: that we can change every aspect of the design, including who wears them and when. I have to encourage them. "This project is much harder than you think," I caution. "Most of you will be in despair at some point. In fact, this is how you will know if you are proceeding in the right way: if you suddenly realize that you have no idea what you are doing." Everyone laughs. They humor me. After all, I am the one with the power; I am grading them.

If I am lucky, though, it really does happen as I theatrically foretell. Everyone feels despair. This despair is a little bit magical. I treasure it! It's the despair of the unknown possibility. This despair helps my students recognize an apparently banal assignment as a real design situation. It's not the same despair as looking down at the number on the scale. It's the despair of throwing the scale away, and recognizing that you just have to give up the numbers, and go to New York, and get some new damn pants.

REFLECTION: HOW IS THE QUANTITATIVE NOT?

The Two Cultures of the Sciences and the Humanities

The topic of my undergraduate honors thesis, which I wrote in 1992, was a debate that emerged in 1960s Britain regarding the relative status of *the two cultures*: the sciences and the humanities. In 1959, the novelist C. P. Snow, who trained as a chemist, argued that humanists should defer to scientists because, after all, science enables Progress—all the conveniences of modern existence. In 1962, responding to Snow, the literary critic F. R. Leavis countered that only the humanities provided insight into life—that which gives our existence meaning.

It seems like an obscure thesis topic: a duel between post-war British intellectuals, conducted via lectures at Cambridge, given three years apart. But the two cultures dispute prefigured a divide at my own institution. At Stanford in the 1990s, students partitioned themselves into two disciplinary camps: techies (science and engineering students) and fuzzies (humanities and social science students). Even as our university, in keeping with American liberal arts traditions, required undergraduates to take courses across the sciences, social sciences, and humanities, the fuzzy/techie dichotomy was pronounced. It indicated not just the major subject you had selected, but something about your engagement with the world and what kind of person you were. From my perspective now, I imagine that the faculty deplored talk of fuzzies and techies, framing the persistence of such ideas as one of those inscrutable undergraduate mysteries. But we did not arrive with those distinctions. We learned them on campus. Indeed, the contrasting terms used for student projects may best encapsulate the fuzzy/techie division. Techie classes assigned *problem sets* while fuzzy classes assigned *papers*. Techies solved problems; fuzzies wrote narratives.

Although the division between techies and fuzzies seemed obvious to everyone, it was another matter to explain this difference systematically and precisely. Although techies might snark that problem sets were more difficult than papers, and fuzzies might snipe that papers required more original thought, neither boast clarified how paper writing and problem-solving were divergent tasks. This apparent impasse intrigued me. If the distinction between techies and fuzzies was so fundamental, why did it resist cogent explanation? So in selecting my thesis topic, I hoped that Snow and Leavis, by methodically setting out their positions on the two cultures, would provide the rational argument that I found missing.

Alas, my goal remained unfulfilled. As I conducted my research, I realized that Snow and Leavis were ideological duplicates, setting forth from the same premises with the same prepositional forms. They both assumed the existence of the two cultures as adversarial poles, and they both imagined inevitable conflict between those poles, conflict that demanded some kind of resolution. Snow and Leavis merely took opposite positions in their imagined battle, just like my undergraduate classmates.

At the same time as I began working on my thesis, I took an introductory computer programming class to fulfill a delayed distribution requirement, and I found that, somewhat to my surprise, writing a program and writing a paper were closely aligned activities. Writing a program, I discovered, was like solving a problem by writing a narrative. Moreover, I realized, writing a paper was *also* like solving a problem by writing a narrative. The only difference was that you couldn't compile and run your papers, so you were never quite sure whether your arguments were sufficiently persuasive. But the process was resoundingly similar. Why, I wondered, was the difference between essay writing and programming emphasized—the idea that the program had to compile and run whereas the essay did not—when all the similarities between essay writing and programming were de-emphasized? Maybe, I thought, if I had realized this earlier, I might have chosen computer science as my major instead of modern thought and literature.

My experience in this story is quite ordinary. The objective has a subjective component, and the interpretive has a problem-solving component: this should not be a revelation. Nonetheless, notions of two cultures and their oppositions persist. For instance, when I became a doctoral student in information science, twelve years after writing my undergraduate thesis, we were schooled on disciplinary contestations such as "Information science: science or social science?" and "Quantitative vs. qualitative research methods: which is better?" Once again, the divisions, and their opposition, were widely presumed.

C. P. Snow's beguiling view of progress encourages these false debates. In this narrative, science enables a better world, and the distinctions that we make between objective (typically quantified) data and subjective (typically qualitative or narrative) data facilitate that forward march. Objective data fits into a story of increasing clarity, control, and prediction, which leads to smooth advancement. Subjective data, on the other hand, bogs us down in quagmires, leaving us indecisive and uncertain.

Human Judgment in Quantitative Measurement

Accordingly, it is a common tendency to focus on the hard, externally objective appearance of quantitative measurements while paying less attention to their semantically soft, subjective underbellies. As an example, let's consider precision and recall, two quantitative measures for the evaluation of information retrieval systems (that is, search engines).[7] *Precision* is a way of describing how many of the retrieved results are relevant to a search query. *Recall* is a way of describing how many of the potentially relevant results were retrieved.

For instance, suppose that you're searching a gardening database using the terms *habitat of squash vine borer (Melittia cucurbitae)*. Ten documents in the database are relevant to that query. When you search, you retrieve a set of fourteen documents. In your set of fourteen results, seven are relevant to your query about squash vine borer habitat. The other seven documents are not relevant—maybe they are about a grapevine borer or about varieties of squash that are resistant to the borer, or maybe they are about spiders or potato chips. Meanwhile, there are three documents that are relevant to squash borer habitat, but you didn't retrieve them. The precision of your search is 7/14 documents that were relevant; the recall of your search is 7/10 documents retrieved.

Precision and recall seem to quantify search engine performance in a straightforward, intuitive way. If I get perfect precision and recall in a search, then all my results are relevant *and* I can be confident that I didn't miss any relevant results.

Beneath the intuitive simplicity of precision and recall, however, lurks a conceptual quagmire. Both precision and recall rely on *relevance*. Relevance also seems to be intuitive and simple. But what is it, really? Furthermore, who determines whether a search result is relevant, and how is such a decision made?

At first, we might assume that relevance is a synonym for aboutness, so that, using our example of squash vine borer, a search result about the borer's habitat would be relevant. But maybe relevance also has to do with what I, the seeker, actually want to do, and with other aspects of my situation. For instance, what if a result is in Arabic and I don't read Arabic? Is that result relevant? Or what if a result describes what was known about the squash vine borer in 1911, but the borer's habitat has expanded into new territories since then? Or what if five of the results say exactly the same thing? Are they all relevant, even if only the first one is helpful?

In a series of publications beginning in 1975, Tefko Saracevic reviewed the ways that information scientists use the term *relevance* and showed that usage is disparate (Saracevic 1975, 2007a, 2007b, 2017). In performing assessments of retrieval systems, relevance is made to invoke a wide variety of related concepts. Precision and recall, rather than being stable and simple, are fundamentally uncertain. They are uncertain because they depend on an assessment of relevance, which is an ambiguous concept that can encompass a diverse range of meanings.

Ambiguity about what relevance means has not, however, hindered the use of precision and recall as measures. Relevance might be a quagmire of uncertainty, but precision and recall remain simple to grasp, report, and plot—as long as we avert our eyes from the instability of their foundation. As Saracevic observed, relevance might be impossible to define, but it has an air of intuitive obviousness. So no one is too bothered by its irresolvable ambiguity. Furthermore, the conceptual fragility of relevance hasn't seemed to cause much harm. I mean, search engines have gotten pretty good, right?

I would argue, however, that the performance of search engines is not as excellent as we tend to assume, and that the conceptual instability of these fundamental metrics matters. Most importantly, when we avoid the complexity of relevance, we underestimate the role of the searcher in search engine performance.

To explain what I mean, let me sketch out the basic process for retrieval evaluation, as first developed by Cyril Cleverdon (1967). First, a test collection is created, and a set of queries (such as *habitat of squash vine borer*) is developed for this collection. (In Cleverdon's day, the test collection would have been thousands of scientific journal articles for a specific technical domain. Today, the test collection might be millions of Web pages spanning a diverse range of topics.) The query might be associated with a brief scenario, such as "A novice gardener in North Carolina wants to know if the zucchini that she has planted will be susceptible to the squash vine borer." For each query, human assessors determine the potential relevance of all documents in the collection. Often, this assessment is performed according to a scale, such as 1–4, with 1 being *not at all relevant* and 4 being *exceedingly relevant*. After this preparatory work has been conducted, an evaluation is performed by running a retrieval system against the test collection to generate results for each query. Precision and recall are then established for each test search according to the assessors' relevance judgments.

The assessors' judgments are therefore key to this entire enterprise. In the information retrieval literature, however, the work of relevance judgment tends to be minimally described, as if it were an obvious procedure. Cleverdon, for instance, briefly explains the 1–4 relevance rubric that was used, but otherwise says tersely that relevance judgments "were decided." Such characterizations make the process of relevance judgment seem like another form of quantitative measurement, where an established scale is straightforwardly applied to each document, just like weighing a person or counting steps.

In practice, however, such assessors are performing remarkably intricate interpretive work. For each query (and scenario, if provided), the assessors must formulate a working understanding of relevance, a concept that we cannot readily explain and find difficult to talk about. They must then tune that working understanding of relevance in the context of the test collection and calibrate it according to the established relevance scale (e.g., 1–4). Each document is then considered against that particular conjunction of query, collection, and scale. Indeed, instead of measuring relevance as some property inherent in documents, it is probably more realistic to think of the assessors as actively contributing to the creation of relevance—contributing, in other words, in a highly significant way to our understanding of retrieval system performance, both specifically (in terms of the relative success of systems being evaluated) and generally (in terms of what *good performance* involves).

A retrieval evaluation is, moreover, an artificially constrained approximation of an actual search situation, in that the relevance judgments are performed in advance. In a real search, for instance, your assessment of one result is typically affected by the other results in the retrieved set; the other results provide another layer of context that contributes to your formulation of relevance.

If we think about all of this invisible work that the assessors are doing, and if we recognize that the information seeker in a real search situation is contributing even more to the construction of search engine *success*, then precision and recall begin to seem both less objective and less robust in their ability to encapsulate retrieval performance.

Still the question remains: does it matter? I think it does. Minimizing the role of the searcher in *constructing* (as opposed to *measuring*) relevance can cause us to misunderstand the results of a system evaluation and to misunderstand what our search engines are able to do. As mentioned in chapter 1's reflection essay, this contributes to a weird, dissonant situation, in which we believe that using search engines is a rote task that requires neither skill nor expertise but at the same time we must contribute significant interpretive resources in order to use search engines effectively (Haider and Sundin 2019). Sometimes we merely give the search engine false credit for work that we, the users, are actually performing (Warner 2009). Other times we fail to recognize that our interpretive labor is required for search engines to work properly, and so we fail to provide that labor. In these situations, we might lazily accept any set of results that generally seem to align with our personal preconceptions about the world, even if those preconceptions have no evidentiary basis, rely on unfounded stereotypes, or are otherwise faulty. Accordingly, a vague and amorphous sense of relevance can make it more difficult to identify and surface situations in which, for instance, search engines reify patterns of systemic racism.[8]

I was talking in this way recently, and my colleague Jaime Arguello, who studies retrieval, disagreed with my characterization. IR researchers are perfectly aware that the metrics they use are only heuristic approximations, he noted. We don't ascribe more importance to our evaluations than they warrant, Jaime continued. We don't think that search engines are actually replicating human thought processes or otherwise creating a system that is truly intelligent, in a human sense. We are not overestimating the capabilities of our systems; we know that there is no comprehension of relevance built into them.

I do not doubt that meticulous and thoughtful researchers like Jaime are very much aware of their work's limitations. But minimizing the ambiguity and complexity of relevance causes the field to move in particular directions, which may diminish, or may seek to eliminate, the seeker's contribution to performance success. Users of search engines, moreover, may assimilate the surrounding rhetoric in unanticipated ways. I am reminded of an intermittent conversation that I have been having with my mother for about fifteen years. "Isn't Google just like magic?" my mom will say.

"No!" I will roar. I'll talk about how Google works and how it sometimes doesn't work that well. Over the years, I've tried really hard to make this conversation succeed, in terms of making Google more comprehensible and less magical. But I always fail. My mother prefers to believe that Google magically understands what she wants, even though this is a wild overstatement of Google's capabilities and doesn't reflect her actual experience with the system.

Computational linguists Emily Bender and Alexander Koller (2020) suggest that a similar propensity to mistake quantitative success for qualitative understanding obtains in the realm of large neural language models, such as Google's BERT, which have recently achieved astounding results on natural language processing (NLP) tasks. But is BERT learning? Bender and Koller argue that models like BERT do not, and cannot, understand language, because they operate outside the realm of meaning. (This doesn't mean that such models can't produce impressive results, but it does mean, as Bender and Koller humorously illustrate, that one probably shouldn't rely on BERT for help when being attacked by a bear.) Bender and Koller observe that most NLP researchers draw a clear distinction between human language use and the operation of models like BERT. Nonetheless, they continue, some academics use misleading words like *comprehension* and *understanding* to describe these models. Imprecise language also appears in the popular press. The layperson may reasonably assume that models like BERT understand an information seeker's intent rather than excel at imitating patterns in training data.

The Conduit Metaphor and the Mathematical Theory of Communication

The work of linguist Michael Reddy (1979) suggests that, in the realm of information processing, it may be particularly easy to mistakenly infer semantic implications from quantitative operations. Reddy argues that English speakers tend to conceptualize the process of communication as the transmission of signals from sender to receiver, making use of what Reddy called the *conduit* metaphor.[9] When we use phrases like *getting ideas across* or *putting ideas into words* or *capturing a feeling* in a sentence we are making it seem as if language is a kind of container that we put thoughts and emotions into, so that all a listener or a reader has to do is extract the thoughts we have placed there. But when we're actually conversing or otherwise interpreting language, we're not merely extracting the sense that someone else has neatly packed into the words: we're constructing sense, often through a process of extended and difficult negotiation between the communicating parties. Of course, just as with search engines, we have always been perfectly aware on some level that listening or reading or otherwise making sense of an utterance involves a lot more work than merely receiving a signal. Often we don't really know what we mean when we communicate, and the responses that we get from others help us start to figure it out. Our reliance on the conduit metaphor, however, makes us less likely to fully acknowledge and respect the work involved in communication, especially when it comes to the receiving side. (The very term *receiving* is misleadingly passive, and more specific terms like *listening* and *reading* tend to be interpreted as passive activities, which is why English speakers can talk about *active listening* but not *active speaking*.)

An unfortunate side effect of our cognitive dependence on the conduit metaphor, Reddy suggests, has been the misapplication of Shannon and Weaver's mathematical theory of communication (MTC) beyond its originating technical context as a means of understanding the transmission of signals from a sending device, across a channel, to a receiving device. MTC provides a way of talking about the probability that the signals that were sent are the signals that were received. For instance, imagine two robots talking on the telephone. The robots are exchanging phrases from an English phrase book, but they don't understand what the phrases mean. The robots are outside, and so there is a lot of wind noise. One robot says "See you Monday," but because of the interference, the other robot hears "See you <garbled> day." Given that the robots were exchanging English phrases, the first robot could have said many possible things, such as "See you Tuesday," "See you today," or "See you someday." (Equally, because the robots were exchanging English phrases, there are many possible utterances that the first robot could not have said, such as "See you mardi," "See you firgly," or "vi ses i morgen.") MTC is a way of quantifying how likely it is that any of those possible things were what the first robot actually said.

MTC doesn't specify the form of the sender, receiver, or channel. Accordingly, MTC could be used to describe the exchange of signals between two computers connected by a network, two robots on the telephone, or two spies writing coded telegrams to each other. MTC, then, provides a mechanism to quantitatively describe any communicative process—as long as we restrict our idea of communication to the exchange of signals without reference to meaning, intent, utility, affect, and other human concerns.[10] The generality of MTC was very exciting to people who were, for instance, working on the first automated retrieval systems in the 1950s and 1960s and who hoped that a unified theory of information exchange would provide theoretical rigor for their work.

The problem here, according to Reddy, is that our default notion of signals, as exemplified by the conduit metaphor, is not arbitrary, meaningless codes but containers of meaning. So especially if we are considering two spies writing telegrams and not two computers over a network or two robots on the telephone, it becomes very easy to talk about communicative intent in terms of MTC, even as communicative intent is supposed to be out of scope for MTC.[11] In conjunction with the generality of MTC, then, the conduit metaphor tempts us to conflate quantitative understanding of signal patterns with qualitative understanding of meaning. This in turn leads us to focus on signals (data) rather than the human interpretive processes that people use when they send and receive those signals. Focusing on signals is enticing for all the reasons that retrieval researchers are happy to bracket out the human messiness of relevance judgments; improving quantitative metrics is a much more reliable enterprise than roiling in sticky quagmires of meaning. We can likewise see this slippage of MTC into the semantic realm as an extension of two cultures thinking. Instead of a duality between signals and semantics, which MTC actually suggests, our predispositions toward a two-cultures dichotomy encourage us to choose one instead of the other.

Even at the time that Reddy was writing in 1979, MTC was falling into disfavor as a general theory of information, broadly construed (Machlup and

Mansfeld 1983). But many fields continue to pursue quantitative manipulation of formal patterns as a substitute for the interrogation of semantics; this is what machine learning techniques are all about (the language models that Bender and Koller criticize are but one example). A common endorsement of such techniques is that they that enable data to *speak for itself* in an objective manner. In contrast, when people interpret data, they risk contaminating it with subjective bias.

Objectivity as That Which Speaks for Itself

The notion that human judgment pollutes scientific attempts to understand natural phenomena *as they really are* may seem like a stable and uncontroversial value; however, as Lorraine Daston and Peter Galison (2007) have established, objectivity is a fairly recent historical development. In Daston and Galison's account, which focuses on scientific visualization, objectivity arose in the nineteenth century, congruent with the development of photography. Before photography, scientific illustration attempted to portray an ideal exemplar rather than an actually existing specimen. In other words, instead of drawing a realistic portrait of an individual fruit fly—which has unique, idiosyncratic characteristics—an eighteenth-century scientific illustrator drew an ideal fruit fly. This ideal representation would better portray average fruit fly characteristics, even as no actual fruit fly is ever perfectly average. (To refer to the adventure essay in this chapter, drawing a perfect, or perfectly representative, fruit fly instead of any imperfect, abnormal specimen is similar to reporting one's "real" weight as opposed to what might have appeared on the scale at any individual time point.)

With the advent of photography, drawings of ideal types began to lose favor. The machinic eye of the lens was seen as enabling nature to *speak for itself*, providing access to a truer, more objective reality than the human eye of the illustrator. Daston and Galison emphasize, however, that this initial confidence in the pure eye of the machine was swiftly undermined. Scientists soon realized that photographic devices introduce their own distortions into the images that they produce, and that no eye provides an unmediated view onto nature. From the perspective of scientific visualization, the idea that machines allow us to see true has long been outmoded. In everyday discourse, however, there is a continuing tendency to characterize the objective as that which speaks for itself without the interference of human perception, interpretation, judgment, and so on.

This everyday definition of objectivity particularly affects our understanding of data collection. If in our daily lives we tend to overlook the diverse, situationally textured sense-making actions that information seekers, conversation listeners, and other recipients of communicative acts perform to make automated information systems function, we are even less likely to acknowledge and value the interpretive work of data collectors, even as these actions create the conditions of possibility upon which data analysis can operate.

Our propensity to lose track of the diverse set of interpretive judgments packed into every instance of data collection, and accordingly to diminish the socially situated conditions in which data is created, extends even where data collection appears tightly controlled. Indeed, the interpretive flexibility that pervades data collection has been especially well described in the sciences. Scholars have meticulously documented the sociotechnical processes by which the context of observation is variously assumed, accounted for, forgotten, and reconstructed in the collection, aggregation, and use of scientific data.[12] To summarize what such studies show, here's a brief scenario based on the example of smartphones from the adventure essay. Let's imagine that we're a team of scientific researchers conducting a Movement Census project. To determine how much the residents in our area move every day, we're collecting step counts and distance traveled from a diverse set of smartphone users over a period of two weeks. We know that different phones produce different results, so we make sure to document the hardware and software for each study participant. We also know that gaits vary, so we instruct participants to select from three gait styles: smooth, bouncy, and semi-bouncy. Subsequently, we develop a normalization function to equate data for different devices and gaits. Our function performs pretty well: it can account for 80 percent of the variance between phones. We only have resources to test our function on three popular models of Android phones, but the majority of smartphone users have Android phones. Of course, we'll summarize these limitations in any academic publication that arises from our analysis. We're responsible scientists.

Over time, however, we disregard our pledge. We gradually forget that our attempt to account for variation between devices and gaits was only partial, not complete. Moreover, we do not fully comprehend the particularities and qualifications that inhere within our dataset. It's entirely likely, for instance, that some participants had difficulty selecting a single gait style (bouncy or semi-bouncy?) but we, the researchers, didn't provide a way to select multiple styles or to indicate uncertainty in a selection. Furthermore, our ideas about gait didn't account for people with physical disabilities or infirmities, who might move differently or use different kinds of prosthetics or supports. Indeed, I could go on and on about the tremendous array of decisions that our Movement Census team made in shaping this very particular dataset, including, of course, the initial idea that step counts are a good proxy for movement. In summary, the quantitative data of step counts arises from a complex and intricate array of interpretive decisions, from the way that we designed our study to the individual actions of the contributing participants. Empirical studies of science invariably show similar conclusions.

The Movement Census scenario represents typical practice, not bad science. The problem, if there is one, does not lie with sloppy data collectors; it lies with our continued reliance on two-cultures dichotomies, in which objectivity and subjectivity can be neatly separated and human messiness can somehow be avoided in data collection performed by humans (or with automated devices created by humans). When we imagine that datasets of properties like step counts speak for themselves, we negate the responsibility we hold for determining which properties will be expressed as data, in what form, and with what parameters. To refer to another example from this chapter's adventure essay, it's not any more or less speaking for themselves to count and weigh the apples in the fruit bowl rather than taking a bite and describing the experience. As properties, quantity and weight are not more "appley" than taste and texture.[13]

Nonetheless, despite the undeniably consistent picture that we see across studies of scientific data collection, the desire to remove the human from the data in order to enhance objectivity remains very strong. Invariably, it seems like the ethical move. My students, for instance, very much want to *let things speak for themselves* when they approach a project like designing a schema to describe a set of things. What my students paradoxically fail to realize, in their zeal to be responsible, is that describing things by certain characteristics rather than others merely because those characteristics are countable is a profoundly subjective decision. I remember vividly one especially conscientious student who designed a schema for describing socks. To keep her data as objective as possible, she specified only quantitatively measurable attributes, such as thickness in millimeters, circumference of the ankle opening, and precise composition of materials. She avoided anything that had the appearance of human judgment, such as what the socks might feel like on human skin, what outfits they might complement, or their stylishness.

But was her data objective? Not at all. The circumference of the ankle opening? That's one of the most subjective data elements I could have possibly imagined. What a useless bit of data! It was selected solely because the data creator had a personal preference for the appearance of objectivity. When we view objectivity and subjectivity as opposites rather than complements, this is the kind of trap we find ourselves falling into.

This two-cultures thinking, moreover, distorts the empirical realities of data collection, the challenging work of forcing unruly phenomena to *speak* in clean, distinct, ideally quantitative phrases. It is likely, for instance, that the designer of the sock schema considered the actual measuring of a sock's ankle opening to be unskilled drudgery, something anyone could do. But even the bare mechanics of measuring a floppy circle are tricky. And there are ontological complexities also. Are we measuring socks as unique material items (for every sock in the world, new or worn, a measurement) or are we measuring socks as a class of equivalent copies (one measurement for a set of equivalent socks, e.g., a particular brand and type)? Even sock measuring is not a mindless task. (Issues of versions and exemplars are more substantially addressed in chapter 3.)

This puts us into a situation similar to that of information seeking. Even as human semantic labor is a necessary component of search performance, we don't tend to value this labor. If a search seems to work, then it's because of Google magic. If a search misleads, then perhaps we will rally for increased information literacy, but only as a remedy to technical failure and not as a valuable component of a sociotechnical process. In the case of data collection, we may proclaim that automated systems are only as good as the data that powers those systems. Nonetheless, the actual work of creating data is not something that most of us think much about.

The Human Work of Data Collection

In an advanced elective that I teach (Metadata Architectures), I've put together a project that encourages students to examine the work of data collection. In this project, students undertake three tasks:

1. Collectively develop implementation guidelines for collecting data according to an established standard.
2. Collect some sample data.
3. Review their mutually created dataset and write about what they find there.

The students collect data about video games.[14] (What do they find in their assembled dataset? Ambiguity and dissent.[15]) Initially, many students are skeptical about this project. In particular, they imagine that there is nothing to be learned from the process of generating data, because data collection is the mere recording of objects speaking for themselves. Performing the work, however, reveals this supposition to be false. People make objects speak as data.

In a small way, the data collection project requires the same kind of skill that I first noticed as an undergraduate programmer: solving a problem by writing a narrative. The data that a student creates for any particular game instantiates a unique set of relations between logical requirements and interpretive judgments. Price, for instance; if there is one price for the currently available streaming version of a game and another price for an individual set of unopened, shrink-wrapped CDs from the original Japanese release in 2000, then selecting one or the other or both prices inscribes radically different relations between the accompanying data elements. Depending on which price or prices you use, you concurrently suggest that the rest of the data you are creating for that game should fit one of three possibilities:

- Whatever you get when you stream the game right now.
- A unique physical copy of the original Japanese game from 2000.
- Both #1 and #2 and perhaps additional versions as well.

In this manner, an interpretive judgment about one data element (like Price) entails a logical restraint on the other elements (such as Creation Date or Language), which subsequent interpretive judgments must accommodate or risk incoherence. It's kind of like planning a journey: there are infinite routes, each with certain qualities (the scenic walking route that goes past the bakery with excellent sourdough or the fast driving route that requires a blind merge at speed), but as you make initial decisions, the set of options narrows. There's a logical component—you have to get to your destination—but the experiential texture of the trip is also at issue.

In designing this project, I had hoped to inspire such thinking in my students. But it has not turned out that way. The Metadata Architectures students experience data collection as intricate and meticulous labor, but they don't value their work as something to be proud of. The last time that I taught Metadata Architectures, a student helped me to understand why. It was incredibly frustrating, the student ranted on a class discussion forum, to spend so much energy on a task that almost no one would appreciate. It was like being asked to paint a masterpiece at nanoscale, so that it could only be viewed with a microscope. But could it be, the student wondered, that his frustration also had to do with the low status generally accorded to this kind of work? Employers, for example, prize data analysis, and avidly recruit data scientists with high salaries. Data collection, on the other hand, is outsourced at poverty pay, well beneath professional status. Could the fact that data collection is widely perceived as unskilled and mechanical make it hard to like, even as one discovers that this perception is erroneous? In fact, could the revelation that data collection requires skill and finesse actually make it more unpleasant to perform, given the lack of respect and remuneration?[16]

The student was right. We can't expect better data if we treat data collection as a task beneath our notice, to be performed by machines or low-wage gig workers. Indeed, this has always been the not-so-secret heart of the two-cultures debate, from Snow and Leavis to today: prestige, power, and money. (In Snow's day, programming was the low-status job, seen as relatively mindless and often relegated to women.[17]) For many of us, breaking down the two-cultures dichotomy, if we want to be real about it, will involve giving something up, even if that something is merely a sense of our superiority and specialness, in our aspirations to be data scientists (sexy!) as opposed to data collectors (not even dignified with a real job title). The real revolution in data labor will be in acknowledging that data collection should be celebrated for its skill and creativity, and not just endured in its drudgery.

3
EQUIVALENCE

This chapter contemplates assessment of functional equivalence, by which two similar things might be designated as two copies of the same thing or as two different things entirely. Although we perform such assessments often in our daily lives, we tend to do so without much conscious thought. I surface equivalence decisions as a fundamental component of data integrity.

In the adventure essay, I tell stories about

- Lurpak butter in Denmark and Lurpak butter in the United States.
- Thin Mints in California and Thin Mints in Texas.
- American Coke and Mexican Coke.
- Editions of *Hamlet* in North American library catalogs.
- Editions of *Mr Jelly's Business* in the Austlit catalog.

In the reflection, I ruminate over

- What *one thing* might constitute.
- Disciplinary assumptions about information things (documents).
- Decisions about things in data design and implementation.

ADVENTURE: WHEN IS BUTTER?

Lurpak Butter in Denmark and Lurpak Butter in the United States

One morning in Copenhagen, I looked down at the butter dish and realized it was almost empty. Hadn't we just bought butter? I wondered. Why did the butter in Denmark disappear so much more quickly than the butter back home? Danish butter *is* delicious, but European butter is available in a lot of American supermarkets also. Once or twice, I'd even bought the exact same Lurpak butter in North Carolina, and I didn't remember using it up so fast then. Did I have to start watching my butter intake? I frowned, and made a note to replenish the butter on the way home that evening.

When I was unwrapping the new butter the next morning, I noticed something. The butter in my hand was 200 grams. But in the United States, butter is sold in ounces, not grams.[1] Sure enough, the Lurpak butter that I might buy in North Carolina is 227 grams (8 ounces). It's not the same butter! The Danish butter is a wee bit smaller.

Except that it *is* the same butter. I mean, the butter *itself* is the same. It's just that the size is slightly different. It's like the mustard that I buy. My local supermarket in the United States stocks Maille Dijon mustard in small jars (7.5 ounces, or 213 grams). But in Denmark, the Maille jars are much bigger (380 grams). It's the same mustard. The jar even looks the same, it's just that the American one is small and the Danish one is big. Well, of course the language on the packaging is different. That's true of the butter as well.

So it's not the same butter? Or it is the same butter?

When *are* two things the same thing?

Of course, no two things in the universe share the same existence. Believe me, I know; I have a twin sister (see figure 3.1). Which is which in the photo? I have no idea. But that doesn't make us equivalent people.

Figure 3.1
My twin sister and I were almost indistinguishable as babies, but we're not functionally equivalent. (Photo by Karl Mangoian.)

With many kinds of objects, though, we treat copies at various levels of fidelity as functional equivalents. Some packages of butter—for instance, the 200 gram packages of salted Lurpak butter on sale in Denmark—can substitute for one another interchangeably. Other packages of butter, such as the 200 gram packages of salted Lurpak butter that I might buy in another European country, are almost exact copies of the Danish ones: only the language on the packaging differs. And there are other versions of Lurpak butter, such as those that I might buy in the United States, that are pretty darn close to the European ones, except maybe just a smidge larger or some other negligible difference.

When I ask if the butter is the same or not, I'm asking about this kind of practical, everyday equivalence. In our daily lives, when we talk about *things*, we are often referring to classes or types (such as *salted Lurpak butter*) rather than unique individuals (such as a particular package of butter in my refrigerator). All members of the class share a functional equivalence.[2] We use the language of singular items (things) but we mean a category that might include many interchangeable members (a type).[3]

Moreover, we often use the same term—for instance, *butter*—to refer to classes at varying levels of specificity. Let's say I have some data about the fat content of butter. My data indicates that different butters have different fat content. What is this data actually describing? Is it describing individual packages of butter, like the one in my refrigerator? Unlikely. All the salted Lurpak butter in Denmark, for instance, probably has the same fat content. *Butter* here probably means a particular product sold by a certain manufacturer.

Let's say I have other data about the freshness of butter. What is *this* data actually describing? Is it describing the same butter as the first dataset? All the salted Lurpak butter in Denmark may have the same fat content, but it's not going to be equally fresh, because it was produced and packed at different times. *Butter* here could mean "the subset of a particular product sold by a certain manufacturer that was produced on a particular date." But after it's manufactured and distributed, butter is stored in different ways and for different lengths of time by the people that buy it, so *butter* could mean something even narrower for this dataset. It could even mean individual packages of butter, if the data were limited to, for instance, the current stock of a certain bakery. My fat content dataset and my freshness dataset are describing different classes at different levels of specificity: not the same thing. These datasets can't be directly aggregated.

When we are going about our daily lives, it is easy for us to talk about *things* like butter at multiple levels of abstraction (varying degrees of specificity) and not notice that we are doing so. Here's an ordinary little story:

> We were out of butter, so I went to get some at the market. When I got there, I couldn't find the butter that we usually get. So I got two packages of some other butter. Then, when I got home, I dropped one butter on the floor, and it got a little dented. It's okay. The butter tastes the same!

This story talks about butter at different levels of abstraction: as a basic substance, as a particular product, as a particular product in a specific form, and as a unique item. The level is sometimes ambiguous, as in the last sentence; it might be talking about the dented and undented packages, or it might be talking about the *butter we usually get* and the different product (*other butter*) that was purchased instead. In conversation, we usually don't notice this ambiguity, because it makes sense either way. Still, we can run into difficulties if we mistake the level of abstraction, or if more versions of some *thing* exist than we realize.

Thin Mints in California and Thin Mints in Texas

For instance, in the United States, there is a national youth organization called the Girl Scouts. For many years, Girl Scouts have sold packaged cookies to fund their activities. The cookies come in multiple varieties. Thin Mints are particularly popular and have been sold by Girl Scouts since the 1950s. It's hard to live in the United States and not know about Thin Mints.

People talk about Girl Scout cookies and Thin Mints as if these classes were uniform. But they are not. Nationally, Girl Scout cookies are produced by two different bakeries, and the cookies differ depending on the bakery. The variety that you receive when you buy a box from a Girl Scout depends on which bakery has a contract with the local Girl Scout council. The distribution across the country is vaguely regional—one bakery (ABC) is mostly in the Midwest, while the other (Little Brownie) is mostly elsewhere—but the geographic coverage is inexact. ABC is also in large parts of California and Nevada, for instance. What this means is that when you buy some Thin Mints, your box may contain cookies with a slightly thicker chocolate coating, or your box may contain cookies with a thinner chocolate coating and slightly more prominent holes. (Some people claim that the cookie with the thinner coating has a stronger mint flavor.) When I moved to Austin, Texas, and found some Thin Mints in our office kitchen,

I thought that the cookie recipe had changed, because I was used to the version with the thicker coating. I was correct that the cookies were different, but the cookies had actually not changed at all. What I had assumed was one *thing* was actually two things. In Austin, the local Girl Scouts contracted with ABC, whereas the Girl Scouts in other places I had lived (Los Angeles, San Francisco, Seattle) had contracted with Little Brownie.

In the case of Thin Mints, the differences are subtle and some people don't notice them at all. For other Girl Scout cookie varieties, the differences can be greater, and sometimes the names are different also. The chocolate-covered peanut butter sandwich cookies taste similar, but they have different names: Tagalongs or Peanut Butter Patties. On the other hand, S'mores have the same name, but one version is a graham cookie coated with a layer of marshmallow, and then a final layer of chocolate, while the other cookie is a sandwich of two graham cookies with layers of chocolate and marshmallow on the inside. No one would mistake the two different versions of S'mores as "the same" cookie, tastewise.[4]

Still, if you are traveling around the United States and happen to run into a table of Girl Scouts selling cookies, the scouts will just look at you blankly if you ask them which kind of S'mores they have. For them, there is only one thing: the kind of S'mores that they sell. And in general, when people talk about or think about or buy or eat Girl Scout cookies, whether S'mores or Thin Mints or Tagalongs or Peanut Butter Patties, they do not consider that each variety of cookie has two possible states. Even after I learned that the Thin Mints in Austin were different from the Thin Mints where I had previously lived, I just thought "Oh, someone brought Thin Mints," when I found them again in the office kitchen. I didn't think "Oh, someone brought the ABC Bakery variety of Thin Mints." For most of us, differentiating Girl Scout cookies by bakery would seem pretty strange. On the other hand, for some cookie connoisseurs, even the small differences between Thin Mints matter quite a lot. For those people, it is not at all weird to say that Thin Mints are actually two things. For those people, the difference between Thin Mint varieties seems obvious and distinctive, and lumping all Thin Mints together as if they were the same is a real problem.

American Coke and Mexican Coke

There are lots of similar situations, where some people notice a distinction between two similar things, and other people don't. Once you start taking account of these situations, it's kind of amazing how often they emerge. Also in the United States, there are people who swear that Coca Cola bottled in Mexico is different from—and tastes better than—Coke bottled in the United States. People who prefer Mexican Coke claim that it has a crisper taste. Restaurants in New York can sell Mexican Coke at a markup, as a special item. What makes Mexican Coke different? American Coke is made with high-fructose corn syrup, but Coke in Mexico is made with cane sugar. Also, Mexican Coke comes in single-serving glass bottles, and American Coke is more likely to come in plastic bottles or cans.

In 2011, the food writer J. Kenji López-Alt conducted an elaborate blind taste test where people engaged in multiple comparisons between Mexican and American Coke, with some of the comparisons in plastic cups and some of the comparisons in anonymized cans or bottles. One subset of participants overwhelmingly chose the American Coke; López-Alt suggested that these participants based their selection on taste. Another subset of participants overwhelmingly chose the selection that happened to be in a glass bottle, and López-Alt suggested that these participants based their selection on presentation. Intrigued, López-Alt conducted another test with multiple tasting rounds in which instead of using anonymous labels (Coke A and Coke B) he labeled each beverage as Mexican or American. In some tests he used false labels, so that people tasting beverages labeled *Mexican* and *American* were actually tasting two cups of one Coke variety, either both Mexican or both American. In this second version of the test, people overwhelmingly chose the beverage of their preferred provenance, either Mexican or American.

This Coke example is ordinary. It is common to find situations where distinctions between things such as butter, Girl Scout cookies, and Coke are perceived by a few and not by others. It is also common for people to identify a distinction between things and yet not be able to characterize that distinction with precision. For those of us who do not have strong feelings about the country of origin of our Coke, it seems bizarre that people would make such a fuss about a minor difference. It therefore seems satisfying to discover that people who believe strongly in a distinction—that there are two things, rather than one—are mistaken when they attempt to describe the basis of that distinction. Accordingly, when the results of López-Alt's test are revealed, the first reaction of someone who has not previously considered Coke provenance to be of undue concern is likely to be schadenfreude. See! There is no distinction to be made here. People who claim to prefer Mexican Coke are delusional, and they need to acknowledge the actual data and believe the evidence.

But this is an incomplete conclusion to draw. There are indeed clear distinctions to be made between Mexican Coke and American Coke: in taste, in presentation, and in one's preference for a particular provenance (even if this preference is not actually due to taste, it is nonetheless in practice a clear differentiating characteristic). When the characteristics are presented separately, these distinctions are made with relative consistency. In everyday life, though, it is difficult to disentangle these characteristics. We are apt to say merely that Mexican Coke is *better* without precisely identifying the factors that contribute to this decision and tracing how these factors relate. As a result, we tend to assume that taste and provenance are working as a single characteristic of differentiation: Mexican Coke is better because it tastes better. Actually, though, taste and provenance are two separate characteristics that operate independently. The problem is not in perceiving distinctions that don't exist but in being unable to sort through a complex ecosystem of characteristics to identify which ones matter and how they do so.

This kind of situation is pervasive. We tend to find it difficult to identify the actual characteristics upon which we make distinctions, even as we maintain that there is a clear distinction to be made. As with the Coke test, this often occurs because there are multiple principles of differentiation at work, and we have not thought about how these principles matter, in terms of establishing whether we have two *things* or one *thing*.

In philosophy, a vocabulary for talking about this phenomenon is by *accidental* and *essential* characteristics. Essential characteristics are the more important ones, necessary for two specimens to be part of the same species; for instance, for two organisms to be two human beings instead of one human being and one centipede or one human being and one robot. *Accidental* here does not mean that a characteristic is unintended but that it is not vital to identity. For instance, most people have two legs, but having two legs is not an essential characteristic of humans—you can have one leg or no legs or really any number of legs and still be human. Or . . . well, what if you do have six or eight or one hundred legs? Is that too many legs to still be human? Talking about accidental and essential characteristics in an absolute way can be dangerous, because we are likely to be wrong when we imagine that no human will ever have six legs or eight toes or dynamic gender. But ideas of accidental and essential can be useful in a relative, limited way to understand how some characteristics might motivate functional equivalence while other characteristics do not. For instance, the size of the Coke container does not seem to matter in the same way as provenance. We might then call *size* accidental and *provenance* essential—in a relative sense, of course, because we might disagree in our understanding of these properties and their relationship to Coke, or our understanding may change over time or according to certain situations. (We might agree that an American Coke in a small bottle and a large bottle are functionally equivalent in general terms, but if I am thirsty, the size is situationally important.)

If we think back to the butter, we might say that the material composition of the butter itself is essential but that the size of the package and the language of the packaging are accidental. That would make the Lurpak butter that I might buy in Durham, North Carolina, "the same" as the butter that I might buy in Copenhagen. Except that . . . I've kept the butter situation simple so far. For instance, butter might also be differentiated by the addition of salt or the use of organic ingredients or the inclusion of vegetable oil to make it spreadable even when cold. How do salt content, organic ingredients, and consistency affect the functional equivalence of butter?

Oh, for heaven's sake! It all gets absurdly complicated. Cookies, Coke, butter . . . each *thing* explodes into a complex ecology in which the distance from one version to another grows increasingly difficult to chart.

Editions of Hamlet *in North American Library Catalogs*

At one time, I thought that this kind of chaos could be tamed. I thought that I could establish better criteria to determine the equivalence of information objects. Information objects are different from things like Coke, because they have *content*. Think of this essay that you are reading: it consists of a series of symbols (letters, spaces, punctuation) in a certain order. This content can be reproduced over and over: the same content in multiple books, like the same Coke in multiple cans. But when you drink a can of Coke, its contents disappear. When you read a book, its contents remain the same. Even if you've underlined and annotated the book, the content—the printed words on the page—is unchanged. If you've memorized a book, you can hold the contents in your brain as a nontangible kind of copy. Similarly, oral stories might never have been recorded, and yet the stories still exist.[5] In this way, informational objects seem to have an abstract essence, where the existence of the information transcends the material carrier that contains it.

In considering the ecology of versions that might exist for an information object, this abstract nature of informational content has often been a focal point. A notion called the *work* attempts to encapsulate a kind of abstract unity for all versions of an information object. Abstract unity is a grandiloquent term, but the concept of a work is actually simple and intuitive. We use it all the time without realizing it. For instance, if someone asks you to name your favorite play, your response is probably something like "*Hamlet* by Shakespeare," by which you mean any version of *Hamlet*. Maybe your experience with *Hamlet* is reading the Norton critical edition in a college class, or maybe your experience with *Hamlet* is seeing Derek Jacobi's performance in the lead role for British television, or maybe your experience with *Hamlet* is reading a facsimile version of the second quarto available through the Web on Internet Shakespeare, or maybe you've read *Hamlet* in all of these ways and more; you're not thinking of any specific version when you answer the question, just of *Hamlet* in general. That's what the idea of the work is: a way of expressing the fundamental unity that makes all the versions that you might actually read or see combine into a basic sense of *Hamlet*.[6]

Because the work is a concept that we already use—we just haven't recognized what we're doing when we do it—it seems reasonable and natural. Certainly, if I ask you about your favorite play and you respond "*Hamlet*," we both feel like we know what we're talking about. It's because of the apparent simplicity of this notion that I looked at the information ecosystem of copies and versions and thought, ah, here is a problem to be solved.

In 1969, library cataloger Seymour Lubetzky also hoped to solve this problem. Library catalogs, then and now, may have an outward appearance of order, but they don't take account of works, the fundamental unity of information objects. Instead, library catalogs describe information objects at the level of physical manifestations. A book, for instance, is issued by a publisher in a certain edition, such as the annotated edition of *Hamlet* produced by Yale University Press in 2003. This 2003 Yale Press edition may then be released in various physical formats, such as a paperbound book and an electronic book. A library catalog describes each of these two physical manifestations of the 2003 Yale edition separately. If you were to search a library catalog for *Hamlet* by William Shakespeare, you would find separate records for each manifestation of each edition (see figure 3.2).

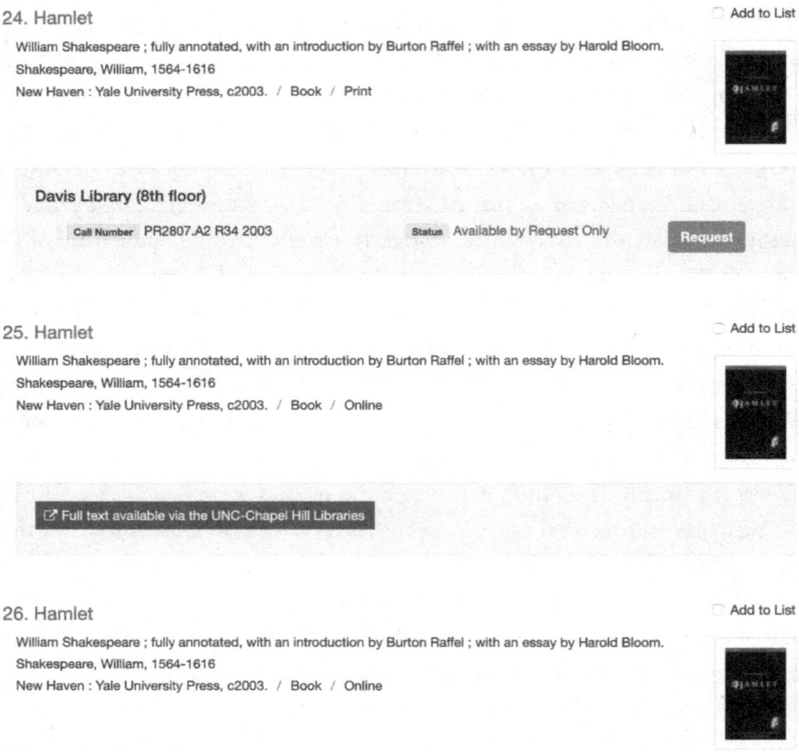

24. Hamlet ☐ Add to List
William Shakespeare ; fully annotated, with an introduction by Burton Raffel ; with an essay by Harold Bloom.
Shakespeare, William, 1564-1616
New Haven : Yale University Press, c2003. / Book / Print

Davis Library (8th floor)
Call Number PR2807.A2 R34 2003   Status Available by Request Only   Request

25. Hamlet ☐ Add to List
William Shakespeare ; fully annotated, with an introduction by Burton Raffel ; with an essay by Harold Bloom.
Shakespeare, William, 1564-1616
New Haven : Yale University Press, c2003. / Book / Online

⧉ Full text available via the UNC-Chapel Hill Libraries

26. Hamlet ☐ Add to List
William Shakespeare ; fully annotated, with an introduction by Burton Raffel ; with an essay by Harold Bloom.
Shakespeare, William, 1564-1616
New Haven : Yale University Press, c2003. / Book / Online

Figure 3.2
If you search the library catalog at the University of North Carolina at Chapel Hill for *Hamlet* (title) by William Shakespeare (author), you'll find 440 results, including 3 entries for different manifestations of the 2003 Yale University Press edition (1 print manifestation and 2 online manifestations).

Actually, you would find even more than this. You would also find manifestations of performances (such as a DVD of a film version) mixed in with print manifestations. And you would find manifestations of Polish translations and manga editions and adaptations into prose and modern language (see figure 3.3). You would find all these because they share a title (*Hamlet*) and an author (Shakespeare).

The set of all of these items (*Hamlet* by Shakespeare) is confusing because its relationships are unclear. In the library catalog list of *Hamlets*, some of the items are very close to each other. The 2003 Yale University Press print version and electronic version share exactly the same words. A 2016

Equivalence

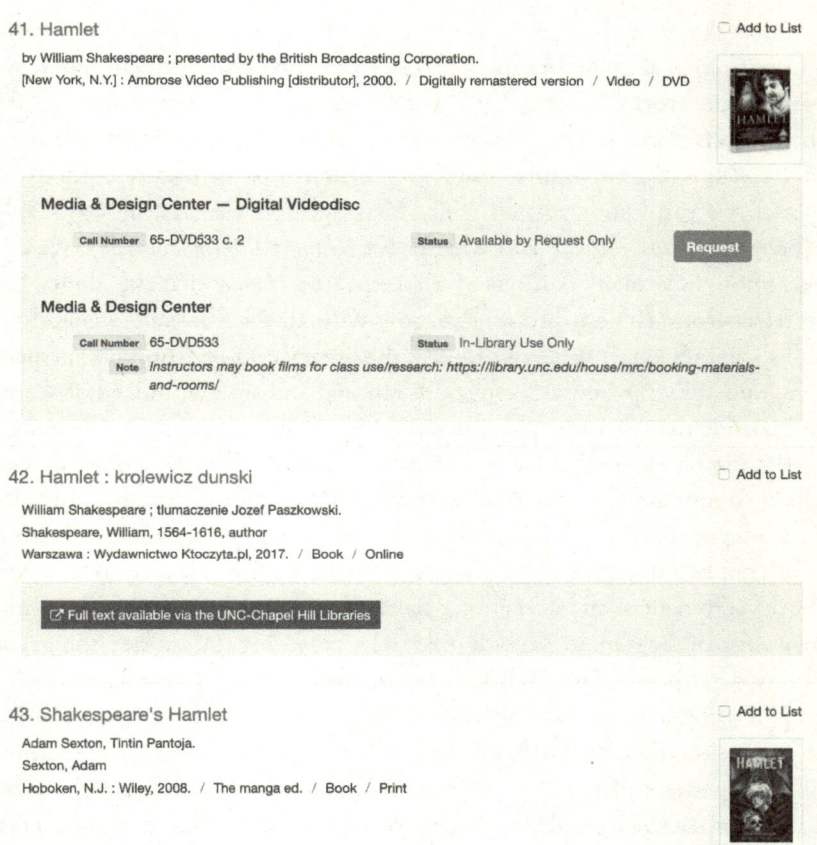

Figure 3.3
Catalog search results for *Hamlet* include a DVD of a 1980 television performance, a Polish translation, and a manga edition.

Bloomsbury Arden edition of the play will have some wording differences, but the content will overall be quite close. And yet some of the items that share this title and author are not so similar. A Finnish film that is a modern satire based on *Hamlet* (*Hamlet Liikemaailmassa*) is very different from both print editions of the play. If it's *based* on *Hamlet*, then it's really not *Hamlet* at all, is it? This is all very similar to our problems with butter and cookies and Coke. The *Hamlet* search results are hard to understand because there are multiple characteristics differentiating the *things* in the list, and it's difficult to extract each characteristic and to trace the extent of difference for each one from item to item.

Lubetzky thought that if we understand that we are describing *works* and not physical manifestations, then we can solve this problem. First, we identify the works: what is and is not *Hamlet*. Then, we group (or *collocate*) the editions of the works. If we make this simple adjustment, then the mess of *Hamlets* will resolve into order. In a system that collocates editions of works, we can know what is really Shakespeare's *Hamlet* (the 2013 Yale University Press edition) and what is not (*Hamlet Liikemaailmassa*). We can also know how many editions of Shakespeare's *Hamlet* that the library has to select from. (It's hard to do that now with all the films and adaptations and so on mixed in.) When we look at the library catalog from this perspective, Lubetzky's proposal seems both rational and simple. Indeed, it seems remarkable that we have not been collocating editions of works all along.

But library catalogs still don't collocate editions of works. Lubetzky was talking about all of this in the 1950s and 1960s, and the catalog continues to display an untidy riot of *Hamlets*. Why hasn't this problem been fixed?

It's not because there isn't a plan. In 1998, Lubetzky's ideas contributed to the development of an international recommendation—Functional Requirements for Bibliographic Records (FRBR)—regarding the entities (things) of library description. The FRBR recommendation posits four related bibliographic entities in levels of increasing specificity, from works (like *Hamlet*) to expressions (like the 2013 Yale University Press edition) to manifestations (like a paperback of the 2013 Yale edition) to items (an actual paperback on a shelf somewhere). A library catalog based on FRBR would, as Lubetzky proposed, show all the expressions and manifestations (editions) that a library might hold for a particular work. However, FRBR has yet to be fully implemented in library systems.

I've read the FRBR report, and once one gets past the technical, fusty-sounding language of *bibliographic entities*, the FRBR model seems both eminently sensible and easily applicable to any information resource. Indeed, at first the set of relationships that FRBR models seemed almost obvious to me, as though anyone could have arrived at a similar plan. How curious, I puzzled, that all information systems designers didn't adopt a similar model. It would solve so many problems! But when I began to think about how I would implement specific examples in the FRBR model, questions and quibbles immediately arose.

Here's an example. FRBR proposes that transpositions between media result in new works. So an opera version of *Hamlet* is a new work in its own right and not an expression of Shakespeare's play. As a general rule, this seems appropriate. But complications ensue. According to examples provided in the report, for instance, staged performances—filmed versions of theatrical productions—would be considered expressions of Shakespeare's *Hamlet* (presumably because *Hamlet* was meant to be performed this way).[7] However, cinematic versions of *Hamlet*—productions conceived as films— would constitute entirely new works. "Franco Zeffirelli's *Hamlet*" is thus distinct from "Shakespeare's *Hamlet*"; these are two separate things and not two versions of the same thing. And yet cinematic versions can be quite traditional in their staging, and theatrical versions can be quite unconventional. Zeffirelli's film of *Hamlet* is arguably much closer to what we know of the original text that an experimental theatrical version set in early twentieth-century Persia.[8]

From my experiences in talking to colleagues and in teaching FRBR to students, I know that my reaction is typical. Everyone who reads the FRBR report wants to change something about it. Then, as one continues to produce examples in which FRBR implementation seems to fail, one begins to think that there must be fundamental flaws in the model itself.[9] After all, the entire point of FRBR is to associate things that are effectively the same, and to separate things that are different. And yet FRBR separates versions of *Hamlet* that are actually quite similar (traditionally staged theatrical and cinematic versions) while equating versions that are quite different (experimentally staged theatrical and cinematic versions). Clearly FRBR doesn't do what it's supposed to do.

Moreover, for each problematic example that one might identify, it's not particularly hard to formulate potential solutions. FRBR might be flawed, one thinks, but a clever person can fix it. We might, for instance, have different sets of rules for different genres. We could have one set of criteria for determining when versions of a play constitute entirely new works, and another set of criteria for poetry, and another set of criteria for technical manuals . . .

As I said, I used to be someone who thought this way. But I no longer think so, and it's not just because a putatively logical and clever approach like developing different rules for different genres would be horrifically complicated. It's because figuring out whether an experimental staging that sets *Hamlet* in imperial Persia is properly a version of *Hamlet*, or whether Persian *Hamlet* is instead an entirely new work, is not a problem that can be solved in the conventional sense. In the realm of English literature, textual scholars have debated these sorts of issues for generations, and they remain contested. Importantly, the focus of scholarly disagreement has shifted over time. What was once unquestioned may now appear open, and what was once open may now appear settled. In some eras, for instance, there was relative agreement about the regulating principles that underlie what *similarity* means for versions of a literary work like *Hamlet*. In the nineteenth and early twentieth centuries, scholars generally thought that the essence of a work was located within its author's intentions, and that the farther any version of *Hamlet* was from what Shakespeare might have envisioned, the less strongly it was associated with the *real Hamlet*.[10] The role of authorial intention in determining similarity between versions was taken for granted as being important, just like we take it for granted that Shakespeare was a great playwright. (Of course, even as scholars agreed that authorial intention was the fulcrum around which a work revolved, they disagreed mightily about what Shakespeare might have envisioned.)

Today, however, textual scholars are skeptical of authorial intention as a regulating principle. Plays like *Hamlet* have long histories of production and transmission, in which printers, editors, actors, and others have made adjustments to the text. Current scholars tend to think of these histories as contributing to the character of a work rather than contaminating the author's vision.[11] If there is any sort of consensus on such issues today, it's that things like *Hamlet*—information things—are inherently plastic. There is no true nature within information objects for us to discover or make progress toward.

Editions of Mr Jelly's Business *in the Austlit Catalog*

What this suggests is that Lubetzky's simple proposal—that the library catalog should collocate editions of works—can never be accomplished to anyone's satisfaction. If we recall the butter, the Girl Scout cookies, and the Coke, this probably seems obvious. There's too much uncertainty about the status of different organizing principles and when they matter in regulating the equivalence of different versions. But I didn't wend my way through a long digression about Lubetzky, library catalogs, and FRBR just to sigh with nostalgia for a simpler time, when we naively thought that grand schemes for universal cyberinfrastructures were merely a matter of will and proper implementation. My admiration for Lubetzky is not nostalgic. It's still a good idea for the library catalog to collocate editions of works, even though collocating editions of works is not a solvable problem. It's still a good idea for all creators of datasets and all creators of systems that make use of datasets to actually think about what *things* are being described and to group those things according to relations of functional equivalence, even though that is also not a solvable problem. We do need to calibrate our expectations for this kind of project and appreciate its limitations. Some of our decisions will be wrong. Other decisions will be wrong for some people in some situations and fine for other people in other situations. As time passes, more decisions will become more and more wrong. But the inevitability of imperfection need not be a deterrent.

Mapping out increasingly intricate rules that attempt to delineate ever more precisely what kind of thing should be functionally equivalent to what other kind of thing at which time for which people—that's ill advised. Increasing the complexity and precision of rules in response to particular conditions—Zeffirelli's *Hamlet* is an expression of Shakespeare's *Hamlet* even though it's a film, but Imperial Persian *Hamlet* is a new work even though it's a theatrical production—might result in some marginal improvements for some people in some ways. But these kinds of efforts also tend to make us delusional, as we become invested in the cleverness of our exquisitely tailored systems. It becomes easy to forget that whatever we're doing is inherently limited—and that for whatever well-reasoned, impeccably logical determination we might make, someone else is going to think that we are absolutely *out of our minds*.

Obviously, organic butter is exactly the same as conventional butter. But ABC and Little Brownie Thin Mints are totally not the same—*that* difference is real. And *clearly*, it only matters if the Coke is in a glass bottle and not whether it's from Mexico or the United States. Of course, when I say it that way, this kind of thinking seems silly and easy to avoid. But it's the things that we don't say—that we don't even realize that we're not saying—that cause problems in the end. Even if we decided that the type of sweetener used in a bottle of Coke was important, for instance, would we think to consider the place of origin of the sweetener? Probably not, because it's absurd and cumbersome and unnecessary and impossible to describe every possible property of things. Except then we might discover that a sweetening agent from a certain point of origin was being adulterated with some cheaper, dangerous substitute. Oh no! Clearly we should have been thinking about the provenance and integrity of ingredients all along, right? We were ignorant before, but now we've made progress: provenance of ingredients, check! That way lies . . . at best, a false sense of security. When it comes to questions of functional equivalence, our fallibility is assured in ways that we simply can't imagine.

But where I used to see a half-empty cup, I've come to believe that half-full is the better choice. Of course models like FRBR are flawed, and it's reasonable to criticize them. Nonetheless, when it comes to relating almost-equivalent *things*, a flawed model can still be pretty awesome. Fully developed implementations of FRBR remain rare, but they exist: one is Austlit, a project of the University of Queensland, the National Library of Australia, and other Australian universities.[12] Austlit catalogs almost a million literary works created by residents of Australia. When you search Austlit—say, for Arthur Upfield's 1937 mystery *Mr Jelly's Business*—you get a list of works of which the novel is the first item. If you select the novel, you see information about all its collocated editions as well as information about its derivative works (such as a 1972 adaptation for Australian television). Because all the manifestations for a single expression are grouped together in Austlit, we can see that the novel was published under a different title (*Murder Down Under*) in the United States, and some UK versions also used this title (see figure 3.4). Despite the different title, these manifestations are associated with a single expression; they are functionally equivalent. In contrast, the editions serialized in Australian newspapers are not equivalent. These are different expressions.

If, instead of Austlit, we were to search the National Library of Australia for *Mr Jelly's Business*, we would find something like our untidy riot of *Hamlets* (see figure 3.5). We don't know that 1943 Doubleday and 1951 Penguin editions of *Murder Down Under* are both functionally equivalent to the 1937 Angus and Robinson edition of *Mr Jelly's Business*. Indeed, we might assume that *Murder Down Under* is a totally different novel entirely. It would be reasonable to think this, because search results often seem wrong or inscrutable, and anomalies are easy to ignore.

The kind of results that we get when we search the National Library of Australia are the kind of results that we get when we search pretty much any collection of heterogeneous, non-unique things: any music streaming service, any recipe database, any Web store or marketplace, or any catalog of anything. Should I look for butter on a major American retailer's Web

Equivalence 77

Figure 3.4
In the Austlit catalog, two different manifestations of the same novel with different titles are linked as functionally equivalent. *Mr Jelly's Business* in Australia is the *same thing* as *Murder Down Under* in the United States.

site, for example, I see a riot of butters even more untidy that the *Hamlets* or *Mr Jelly's* (see figure 3.6).

The butter differs variably by brand (Land O Lakes, Great Value, Kerrygold), packaging (whole or half sticks, tubs, twin packs), amount (total weight), consistency (whipped or not), salt, and composition (nondairy butter made from olive oil, added olive oil, added canola oil). Just as with the library catalog, each *thing* or instance of butter differs from another *thing* in the list in inconsistent ways. There is Land O Lakes salted butter in packages of

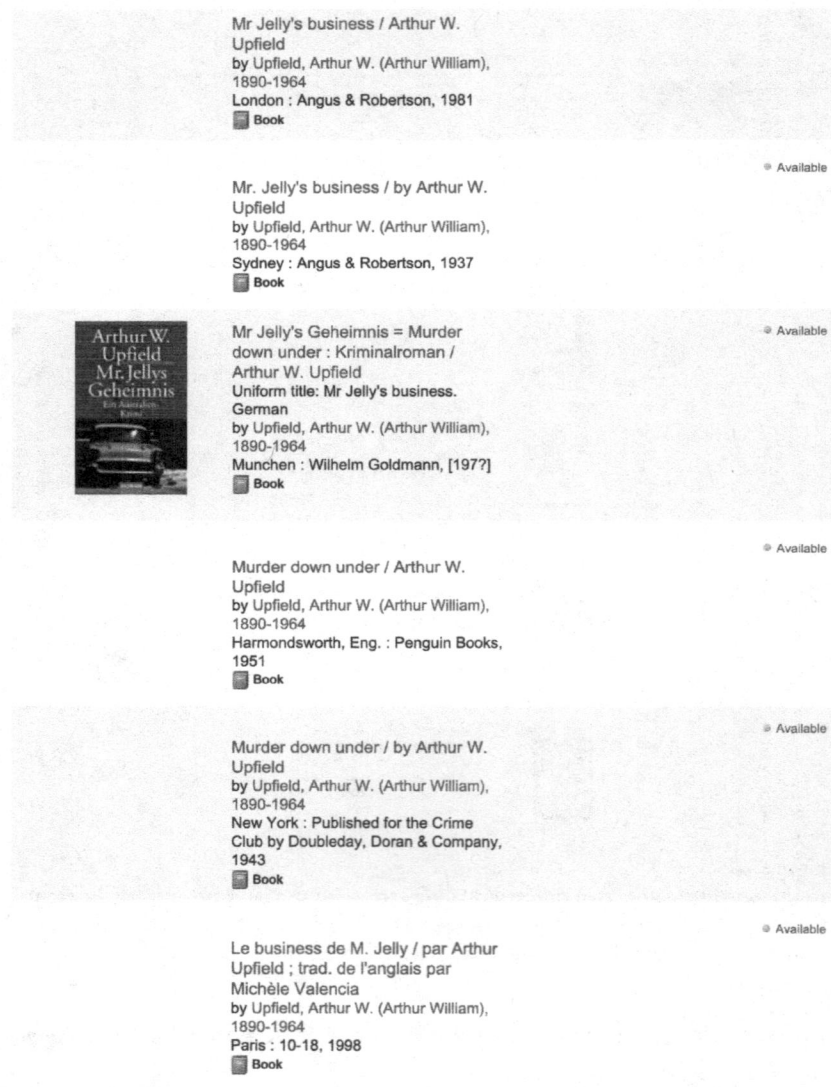

Mr Jelly's business / Arthur W. Upfield
by Upfield, Arthur W. (Arthur William), 1890-1964
London : Angus & Robertson, 1981
Book

○ Available

Mr. Jelly's business / by Arthur W. Upfield
by Upfield, Arthur W. (Arthur William), 1890-1964
Sydney : Angus & Robertson, 1937
Book

○ Available

Mr Jelly's Geheimnis = Murder down under : Kriminalroman / Arthur W. Upfield
Uniform title: Mr Jelly's business. German
by Upfield, Arthur W. (Arthur William), 1890-1964
Munchen : Wilhelm Goldmann, [197?]
Book

○ Available

Murder down under / Arthur W. Upfield
by Upfield, Arthur W. (Arthur William), 1890-1964
Harmondsworth, Eng. : Penguin Books, 1951
Book

○ Available

Murder down under / by Arthur W. Upfield
by Upfield, Arthur W. (Arthur William), 1890-1964
New York : Published for the Crime Club by Doubleday, Doran & Company, 1943
Book

○ Available

Le business de M. Jelly / par Arthur Upfield ; trad. de l'anglais par Michèle Valencia
by Upfield, Arthur W. (Arthur William), 1890-1964
Paris : 10-18, 1998
Book

Figure 3.5
In the catalog for the National Library of Australia, search results for *Mr Jelly's Business* include *Murder Down Under*, but the two titles aren't linked together as functionally equivalent.

Equivalence

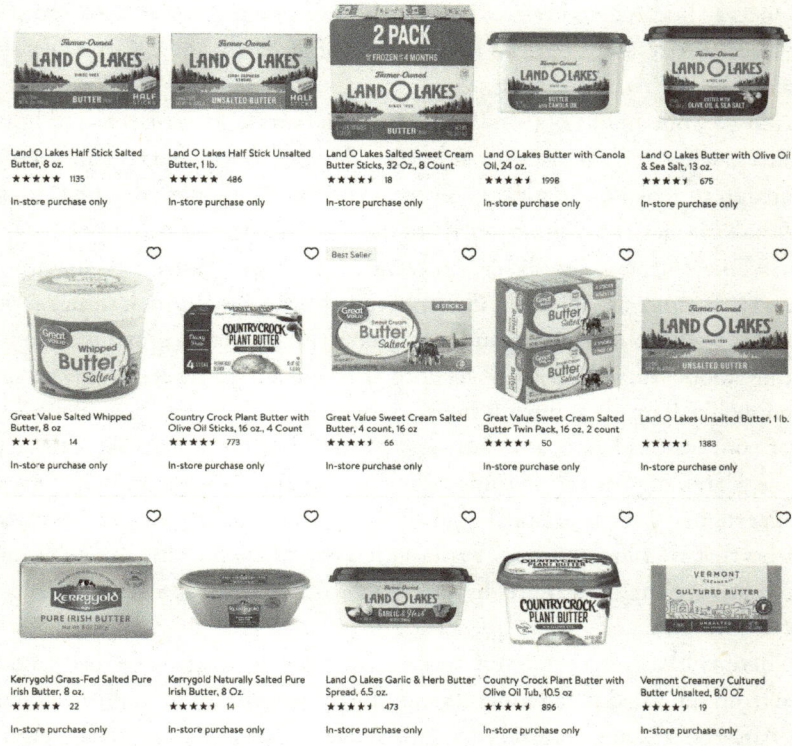

Figure 3.6
A riot of butters, where each item differs from another by an unpredictable set of potential characteristics. This uncertainty around functional equivalence is pervasive in online catalogs of all types, from books to butter.

four sticks (16 ounces total in the package), and Great Value salted butter in packages of four sticks (16 ounces total); there is Land O Lakes salted butter in packages of four "half sticks" (8 ounces total); there is Great Value salted whipped butter in a tub of 8 ounces; there is Land O Lakes "light" butter (with added water) in packages of four sticks (16 ounces total). It's not the same butter! Except when it is.

With the butter, as with the *Hamlets*, so with many—perhaps most—datasets. One *thing* will explode into a riot of vaguely defined units with unclear relations to teach other. Everyday catalogs are particularly dreadful. Online stores, like Amazon? Ghastly. (Are all the versions of different products collocated and related? No.) Online content services, like Spotify or Netflix? Atrocious. (Are all the versions of a certain song or movie collocated and related? Not at all.)

And yet, one might object, if this state of affairs is so pervasive and horrendous, why aren't more people bothered by it? Why hasn't there been a campaign to make our datasets—especially ubiquitous everyday ones such as Amazon—less terrible in terms of understanding the ecology of versions? One reason is simply lack of awareness. It can be hard to diagnose issues with imprecise referents if you haven't learned about such things. I've seen this with students in my university courses. Sometimes, as a lead-in to talking about things and their versions, I might direct students toward an example like figure 3.6 and ask them to explain it to me. At this initial stage before our lessons begin, students tend to talk in terms of the user interface or the search algorithm or some other notion that is more familiar, even as uncertainty with functional equivalence is more salient. In other words, matters that are fundamentally to do with the composition and organization of the underlying dataset are displaced onto other areas. In my experience, this is common.

There is likewise a tendency to minimize the consequences that might arise from muddled data models. From the perspective of an online retailer like Amazon, it's not necessary for customers to make fully informed comparisons of similar versions. Perhaps a customer doesn't realize that there's a slightly newer edition of *Hamlet* very similar to one that they purchased. Is this really such a problem? They found something acceptable, and that's all that matters. But the effects of unclear equivalence are not always trivial. Here's an example: in 2016, aggrieved customers protested excessively high prices for the EpiPen, a patented autoinjection system for epinephrine. (People with life-threatening allergies are often directed to carry an EpiPen or similar dose of epinephrine with them at all times; the EpiPen is the most well known of these injection systems.) As a response, Mylan, the EpiPen manufacturer, produced a generic version of its own product, which was much less expensive. The generic EpiPen was functionally equivalent to the nongeneric version; unlike most generic drugs, it was even made by the same manufacturer. Here, a good understanding of versions and their relations was necessary in order to receive the lower price. If you didn't know that a generic version existed and that it was functionally equivalent to the EpiPen, you'd pay a lot more. The nature of the functional equivalence between products is really important here; if you had a severe peanut allergy, would you want to gamble on a cheaper EpiPen if you weren't certain that it was exactly the same product?[13]

In form, the EpiPen is very similar to the previous, more mundane examples of butter, cookies, Coke, and *Hamlet*. In all these examples, inability to identify and relate versions can result in worse decisions. Sometimes the consequences are negligible. But sometimes the effects are much more serious. And sometimes, the ramifications of such decisions are negligible for some people—for instance, those fortunate enough to have health insurance that covers the cost of an epinephrine autoinjector—but not for others. What's more, the people with the fewest resources—with the least access to information and the least amount of time to negotiate overly complicated systems—will tend be at the greatest disadvantage.

Should we, then, advocate for the Austlit of butter? The Austlit of EpiPens? An Austlit for everything? For sure, the Austlit of butter or anything else wouldn't get us any closer to the truth of whether this butter and that butter are the same. But that doesn't mean that the Austlit of anything wouldn't be valuable. Occasional negotiations with the Austlit of *something* might spur us to focus our attention on matters of equivalence more generally, so that we mark, a little more often, just how information systems can obscure relations rather than elucidate them. When we subsequently face a riot of butters, of *Hamlets*, or of anything else, we can look more closely to the data itself, and we can better sort it out.

REFLECTION: IS THIS THING THE SAME AS THAT ONE?

What One Thing *Might Constitute*

In our everyday interactions with the world around us, we seldom question what things are: where one thing ends and another begins. When I started writing this essay, for instance, I noted four things on my desk: a laptop, two Copenhagen University pens, and an orchid plant. I would have expected anyone else looking at my desk to identify these same four things.

But each of these things on my desk, which all seemed naturally individual to me, can also be seen as collectives. Most man-made things—the laptop, the pens—are collections of parts, and the parts can exist separately from the thing that they help to constitute. If I need a new laptop battery,

I can buy a replacement without needing to buy a whole new computer. Batteries and computers that contain batteries participate as independent *things* in a manufacturer's inventory systems. But after a replacement battery is installed, it becomes part of the computer, even as the potential to remove the battery from inside the computer remains. Similarly, I can disassemble the pens and swap their parts. After this surgery, the pens seem to retain their previous existence as separate, independent things, even as their components have been switched.

Living organisms cannot be disassembled as man-made artifacts can, but they, too, operate as collectives, sometimes in conjunction with inorganic materials. The orchid on my desk can't exist without the soil that it's planted in, and that soil requires a pot. In the same way, the laptop will eventually require access to electricity, typically via a cord, plug, and wall socket, or else it, too, will effectively die and be unusable as a laptop, the thing that we identify it to be.

I, sitting at my desk, also seem autonomous. I perceive a clear boundary between my body and the outside world, and I feel like my body is under my will and control. But even I can be considered a collective. Like every human gut, mine contains a universe of microflora, which functions to maintain my health in ways that medical knowledge only vaguely comprehends. I function as the habitat for a vast microbial ecosystem, even as I have little awareness of this inner universe, and I do not control it like I do my hands and feet. Geneticists associated with the Human Microbiome Project suggest that human beings are best understood as supra-organisms, composites of human and microbial cells (Turnbaugh et al. 2007).

From this perspective, networks and groups are more fundamental in their thingness than individuals. In the adventure essay for this chapter, I referred to my twin, Melissa. Contrary to myth, I don't feel pain when my sister is injured. Nonetheless, our childhood was a kind of joint existence. We were almost never separated, we had words that were intelligible only to us, and we were called "the girls" rather than our own names. In our current lives, my sister and I live in different cities and may not communicate for weeks at a time. Still, neither of us feels alone in the world in the way that many people do. Although it may seem eerie to singletons, this manner of experience is not especially rare. You don't have to be a twin to think in terms of families rather than individual people, forests instead of individual trees, or flocks instead of individual birds—there may be just as much salient identity in these common, everyday collectives as in their constituent parts.

Of course, to examine alternative ways of conceptualizing basic entities like trees and birds—or even more broadly, biological species—does not mean that trees, birds, and species aren't real in the sense of being useful and accurate ontological units. It just means that, as we can see in from the adventure examples of butter, Girl Scout cookies, and *Hamlet*, a single ontological unit is insufficient to contain all of the ways that the world exists. Nonetheless, it can be disorienting to consider forests, flocks, and ecologies rather than trees, birds, and species, because it does imply the human contingency of all such kinds.

As I allude to in the adventure essay for this chapter, all data has an ontological aspect. All data describes some *thing* according to characteristics that align with that kind of *thing*. If we think back to that orchid on my desk, to describe the orchid's level of desiccation and pH level without reference to its soil is to establish the existence of the orchid in a particular way—a way that could cause harmful effects if we imagined the *plant* to be dry even as its soil remained moist. But when we consider the task of collecting data, few of us begin by asking, What is it that I am describing anyway?[14]

If we do consider this question, we are more likely to do so at the level of classes rather than objects. In biological taxonomy, for instance, there are multiple definitions of *species* as a concept, and competing ideas of how the concept of species should be related to other taxons.[15] But the idea of the *specimen* is not debated in the same way. For instance, when a team of evolutionary biologists argued that four species of giraffe should be recognized rather than one species of giraffe, the researchers defined how they understood the notion of a species but not how they perceived the specimens from which they obtained genetic samples (Fennessy et al. 2016).

Disciplinary Assumptions about Information Things (Documents)

Similarly, within information science, the technical literature of knowledge organization has not considered the nature of its specimens to be a central question. The *things* that information systems manage have traditionally been understood as *documents*, that is, carriers of recorded information. As summarized by Michael Buckland (1997), notions of documents have gradually expanded over time to include any sort of information-bearing object, from photographs and podcasts to dead squirrels in a zoological museum and live antelopes in a zoological park. But as regards the process of classification, documents have been assumed rather than problematized.[16]

Instead, knowledge organization has focused on the *content* of documents: their subject matter or *aboutness*. Bibliographic classifications are designed to locate documents that have certain aboutness characteristics. Accordingly, the intellectual project of bibliographic classification has focused on what people write about, with a focus on *knowledge*: that is, the output of scientific endeavor, as typically occurring within academic disciplines. Bibliographic classifications thus enumerate and relate subject categories (such as photosynthesis) as objects of study in the context of their encompassing disciplines (such as botany). The nature of documents has been perceived as incidental to this enterprise.

Intuitively, this position seems reasonable. If we think about a concept like photosynthesis as an element of disciplinary knowledge, the concept of photosynthesis exists independently of being written down or otherwise recorded. So the development of classification schemes based on aboutness doesn't seem to require a theory of documents.

In contrast, as discussed in the adventure essay, library cataloging theorists like Seymour Lubetzky *have* been concerned with the nature of documents. It is, I think, illuminating to see how the two fields of cataloging and knowledge organization, both arising from the same institutional context and both concerned with similar data matters, can approach their work with such different orientations.

To begin this comparison, let's imagine an impressively large nineteenth-century library with thousands upon thousands of titles, all with uniform plain bindings. To maximize storage space, the books have been arranged by size (not an unusual choice, and one occasionally used today). How could you search a collection like this if the librarian were not available to assist? You would have to inspect each book's contents, one at a time. Cumbersome!

With a library catalog, you look through standardized descriptions of books rather than the books themselves. Each catalog entry constitutes a data surrogate of the document that it represents. Instead of inspecting each book, you can go back and forth through the set of catalog entries. If the entries are consistent in content, form, and expression, they are easy to scan and compare. The library catalog, then, is a standardized dataset that enables patrons to identify and select between items in the collection.[17] Much of the technical intricacy of library cataloging involves the application of complicated rules—data collection protocols—to facilitate the systematic creation of reliable data across all the sorts of documents that libraries might hold.[18]

Early catalogs were written as lists, which were challenging to keep updated. Toward the middle of the nineteenth century, libraries began using an individual card for each entry, in combination with physical user interfaces—card cabinets—that supported storage and scanning. This technical advance enabled both easier updating and more sophisticated retrieval capabilities. With a card catalog, it was simple to replicate entries and file them in multiple locations, which supported arrangement by multiple characteristics. The data elements of author, title, and subject became standardized as primary *access points:* the characteristics under which cards might be filed and therefore searched.[19] (Entries included other information such as the publication date, number of pages, publisher, and so on, but these were not access points—in the card interface, they were not searchable.[20])

To summarize, library cataloging focuses on the description of books and other documents in a standardized way—on data creation. The day-to-day work of the cataloger consists of creating entries for new documents, that is, creating new catalog data and adding it to the existing dataset. In this context, it's easy to see how the accumulated experience of creating entries and observing the evolving catalog might cause Seymour Lubetzky to ponder what he was actually describing and to think about relationships between editions and works. In contrast, although Anglo-American cataloging rules have been extended and revised over the decades of their existence, the data that catalogers collect has remained fundamentally consistent in content and structure. Even as, for instance, the terminology of access points is anachronistic in an online environment, authors, titles, and subjects continue to be prioritized as data elements.[21] Moreover, cataloging's data infrastructure—including data creation protocols, vocabularies for data values, and encoding formats—is highly standardized. Ongoing maintenance and design activities are performed by a small group of national institutions or international standards bodies. In the United States, for instance, aboutness data values are typically assigned using standardized vocabularies developed and maintained by the Library of Congress (the Library of Congress Subject Headings, or LCSH). It would be unusual for a library to design its own subject vocabulary.

Knowledge organization, in contrast, is oriented around vocabulary design, and the actual creation of data with those vocabularies is not a great concern. This disciplinary emphasis can likewise be traced to the nineteenth-century library. In that environment, a catalog of data surrogates is one access mechanism for a collection of physical documents. A complementary but otherwise entirely separate access mechanism is to arrange the documents (typically books) in a meaningful sequence, so that related items are in proximity to each other. A bibliographic classification scheme, which groups and relates subject matters within fields and disciplines, provides the structure for such an arrangement. Although such classification schemes support retrieval, their goals orient more toward exploration. Classified collections show relationships between books. The goal is not merely to collocate books about, for example, photosynthesis; it is to place the books about photosynthesis within a more general class of plant cellular biology, within an even more general class of botany, within a more general class of biology, within the disciplinary area of science. The work of the classificationist—the designer of such a system—focuses on the specification of these internal relationships, which will ultimately structure an ordering apparatus, such as a linear shelf order. Accordingly, a classification scheme is often conceptualized as *arrangement*, rather than *description*.[22] One could, of course, arrange a dataset—e.g., catalog entries—in a classified order also. In libraries, however, the older card catalog prioritized direct-access retrieval via an alphabetical (or dictionary) arrangement, so that, for instance, books about photosynthesis would be under *P*, next to books about photography (Cutter 1904). The catalog and classification scheme remained independent access mechanisms.

As the nineteenth century progressed into the twentieth, library catalogs and classification schemes underwent similar trajectories of standardization, with libraries turning to common standards rather than developing unique systems customized for their collections. In this environment, the intellectual heritage of bibliographic classification, with its design-oriented concerns, had little utility. Catalogers, in typical practice, assign books to existing classification schemes; they don't design new schemes.

But classification design did not disappear; it migrated to other contexts. Specialized information services, such as disciplinary indexes for research articles (e.g., Chemical Abstracts or Sociological Abstracts), required specialized classificatory devices—library schemes, created for books, were not sufficiently detailed.[23] In these new environments, the things being ordered by a classification scheme shifted from physical documents to data surrogates (e.g., a citation and abstract for a journal article). Where nineteenth-century classificationists had designed broad, general classification schemes to arrange books on shelves, twentieth-century knowledge organizers designed specialized indexing languages to arrange data surrogates.[24] Initially these data surrogates were assembled and arranged as print indexes. Subsequently they appeared as digital databases. Today such databases—for example, the PubMed database of medical literature or the ERIC database for literature in education—provide full-text access to the articles, but the data surrogates (including index terms that describe the article's aboutness) remain in the background to facilitate retrieval.

Nonetheless, even as the knowledge organizers created domain-specific thesauri to support the retrieval of journal articles from research databases, they retained a conceptual orientation similar to their precursors, the bibliographic classificationists.[25] They saw themselves as creating systems of relations: as creating the conceptual infrastructure to connect photosynthesis to plant cellular biology to botany. As designers, they tended to assume that if this underlying conceptual infrastructure was designed with elegance and precision, then its application—the assignment of index terms to actual documents—should be trivial. Accordingly, the technical literature of knowledge organization focuses on design of data infrastructure and not on the use of that infrastructure to create data. Actual data creation—whether assigning a bibliographic classification number to a physical book, or assigning thesaurus terms to inventory items in a clothing retailer's online catalog, or selecting a code from the International Classification of Diseases to indicate the primary cause of death on a death certificate—is ancillary to the design task.

Decisions about Things in Data Design and Implementation

When I became a doctoral student studying knowledge organization, I knew nothing of the disciplinary evolution that I've described in the preceding section. But the design orientation aligned with my professional experience in San Francisco Internet companies. Before graduate school, I was a content strategist—a fancy title for what was essentially classificatory work. At the consulting company where I worked, design (*strategy*) was valorized. I remember how, at one point, company management proposed to divide our role into two separate positions: content engineers and content strategists. Content engineers would concentrate on implementation, rather than design. But none of us wanted to be content engineers. Everyone claimed—and really believed—that their personal skills tilted strategic. We had all been acculturated to a corporate environment in which *strategy* was accorded power, prestige, and glamor, while *implementation* was expendable and potentially outsourced.[26] In graduate school, therefore, I did not question the focus of knowledge organization. I was happy to concentrate on the design of vocabulary systems as a kind of self-contained project, with only an incidental relationship to practices of actual data collection and to the *things* that a vocabulary might describe. This orientation aligned well with my existing self-concept as a strategic thinker.

In consequence, it took years for me to fully assimilate that classifying something is just creating data about it, and that, however intricately structured with internal relationships, a classification scheme (or indexing language) is at its basis a set of possible data values. Likewise, it took a long time to realize that although a vocabulary like the International Classification of Diseases (ICD) does establish the conditions of possibility for data creation—if it doesn't exist in the ICD, then you can't die of it or be diagnosed with it, datawise—only data creators interact with the ICD itself. Actual users interact with data that has been structured with the ICD: death certificates, electronic health records, and so on (Bowker and Star 1999). The designer of data infrastructure, in other words, shapes the world only at a remove. It is the data collectors, who apply the data infrastructure, the implementers—the medical coders applying the ICD, the corporate taxonomists applying their company thesaurus, the library catalogers creating a call number with the Library of Congress Classification—who really shape the world. This may seem very obvious, but when someone is designing a complex system like the ICD, it is easy to think that that system is the design product, even as the real end products are the many datasets that use ICD codes to describe diagnoses or causes of death and so on.

It is somewhat embarrassing to admit, but I assimilated the full force of this realization only when I became a professor and began to teach my own classes. The first time that I taught an introductory course in organizing information, I planned two related projects: first, design a system (or schema) for describing a set of things; and second, take one of the schema attributes and develop a taxonomic structure for its values. This project would demonstrate empirically that classification is a particular form of description and that describing something is collecting data about it. Students would design their schemas and taxonomies to describe any set of things they might be interested in: science fiction television series, taquerias, yoga poses, natural hair care products. Although my syllabus described these projects in some detail, at the beginning I didn't include much about defining *a set of things*. It seemed like a minor detail.

But issues with thingness arose for almost every project. A student describing cafes, for instance, seemed to conflate a cafe as a *business* (which may have multiple locations) with a cafe as a *place* (an individual location). A student describing energy bars seemed confused about whether a *bar* was a brand, a product within a brand, or a flavor within a product line. Concurrently, it was challenging for students to understand these issues: distinctions between cafe-as-business and cafe-as-location were difficult to absorb. It was, I began to understand, an uncommon and almost unnatural task to pinpoint thingness in a systematic way, for the purposes of data collection. My students did not have the conceptual tools to comprehend this task.

It's not a surprise that students were befuddled, because practical guidance for clarifying relations between things and their data, when it exists, can be rather tortuous. Take, for instance, a general guideline for data creation promulgated by the Dublin Core Metadata Initiative: the 1:1 principle.[27] The 1:1 principle specifies that any descriptive statement (any attribute/value pair, or any data) shall apply to a single resource—a single informational thing. To see why the 1:1 principle is tricky, consider the data presented in table 3.1, and imagine that this data is with associated an image of bombed-out ancient ruins in Syria.

Most people assume that the data given in table 3.1 relates to one thing: a photograph of destroyed ruins in Palmyra. But this data describes (at least) three things at once. Some of the data values relate to the event depicted in the image; some of the data values relate to the image as it was taken; some of the data values relate to the image as it was uploaded and stored in some new system. This data violates the 1:1 principle all over the place.

For most people, however, the ambiguity in table 3.1 doesn't seem like a problem. In fact, it's restructuring the table to remove the ambiguity and satisfy the 1:1 principle that is confusing. This is why, as Richard Urban (2014) reported, data produced by cultural heritage institutions (libraries, archives,

Table 3.1
Example of typical data that might be associated with an image

Data Attribute	Data Value
Photographer	Ixchel Ortega
Event	Destruction of Decumanus Maximus, Palmyra, Syria
Event Type	Bombing
Actor	ISIS
Actor	Syrian Army
Event Date	2015
Source Date	April 4, 2016
Upload Date	April 26, 2016
Source	Press Wire Service
Format	JPEG
File name	Decumanus example.jpg

and museums) consistently violates the 1:1 principle. Urban argues that these violations do not occur because cultural heritage professionals are sloppy and careless; they occur because implementing the 1:1 principle adds layers of complexity that appear unnecessary. It is, for instance, a violation of the 1:1 principle to associate Leonardo da Vinci as the creator of a digitized file of the Mona Lisa. (Leonardo created the painting that the digitized file represents; someone else created the digital image.) But it's weird for people to see an image of the Mona Lisa with only the name of the digitizer as the creator of the image. It seems wrong.

According to the perspective behind the 1:1 principle, human perceptions of weirdness or wrongness are beside the point. The goal of the 1:1 principle is not to facilitate human understanding but to enable automated processing of aggregated datasets in which originating human norms might be various and unknown. From this perspective, Urban's observations miss the forest for the trees. For instance, just because data is structured according to the 1:1 principle doesn't mean that the data needs to be displayed that way. One could design display interfaces so that data for multiple, related things could appear together in any configuration—so that all the data about the Mona Lisa was merged together on one screen even as data about the painting and its digitization remained structurally separate.

Moreover, problems with data referents arise in the absence of specific action: as time passes, or as data migrates outside of its originating context or is aggregated with data from elsewhere. Quietly, ambiguities are compounded and then start to resemble inaccuracies, even as the original data was perfectly correct. For instance, the image that table 3.1 refers to depicts only rubble, no people. The image documents the aftermath of an event in which ISIS and the Syrian Army were actors. For the creators of table 3.1's data, it might seem very clear that *actors* participate in *events* that motivate images and that the relationship between actor and image is causal. But an outsider looking at the data might reasonably connect *actors* to the image rather than the precipitating event, assuming incorrectly that actors appear in the image. Computational techniques can accelerate and intensify these effects.

So, the 1:1 principle makes a lot of sense as a best practice for data design. Nonetheless, as Urban's research indicates and my own experience with students likewise suggests, the 1:1 principle might not be worth the trouble. Even for the limited purposes of a specific dataset, people find it alien and uncomfortable to specify things in a precise and regimented way.[28] It is easier to live with referential ambiguities than to sort them out. Indeed, the whole enterprise of sorting them out is fundamentally a machinic kind of operation, one that people can perform only contingently and provisionally. (It is perhaps emblematic of machinic limitations that machines need people to perform such tasks, even as these tasks seem unnatural and largely unnecessary for people in the absence of machines.)

In some ways, then, my efforts to render the infrastructure of thingness more open and visible for students have convinced me that things will always remain somewhat in the shadows—that we can never know them in a stable or satisfying way. Perhaps paradoxically, I am likewise convinced that thingness remains fundamental to any concept of data, and that we must continue to grapple with it, consciously and critically, even as it constantly slips out of our hands. Why? Because data must always describe some *thing*, and things are elusive and stubborn, like unconscious habits. My point here is not merely that things can always be defined differently but that the contingency of things is so persistently hard to see. We can't decide that collocating versions (of *Hamlet*, of butter, of anything) isn't worth the effort if we don't understand what's going on with the *Hamlets* and the butter in the first place. Even more importantly, in a world where well-meaning people are convinced that looking to *the data* will save humanity from our worst impulses, we need to acknowledge that *the data* exists only in relation to human conceptions of *things*—conceptions that are no better than partially accurate part of the time.

Consider datafication, the process by which people are translated into digital profiles as our online activities are recorded, collected, and aggregated—where we go, what we look at, what we search for, and so on. But is my online activity really data about *me*? My sister and I, for example, often buy presents together for family members, and we switch off who places orders, although not in a predictable or systematic way. A lot of my Amazon purchases, in other words, are really about an *us*. Similarly, my Netflix viewing comprises an evolving set of negotiated agreements with my partner Jason, which includes unmarked distinctions such as "shows that we can watch without one of us complaining too much" and "shows that one of us will watch after dinner while the other one falls asleep on the watcher's lap." These kinds of situations aren't unusual; they are typical. But datafication tends to be discussed as a process of individuation, resulting in digital twins or doppelgangers or other surrogates for people as individuals, rather than networks, collectives, or other configurations of thingness.[29] Anyone's understanding of *me* based on my Netflix viewing or Amazon purchases isn't exactly wrong, but neither is it right. And, well, of course it isn't! This shouldn't be a surprise to anyone at all, because we all have similar experiences, and all of our online interactions can be read in terms of different underlying things.

But we *are* surprised at such realizations, with depressing regularity. Indeed, the account just given—of my fumbling recognition that thingness is fundamental—is likely, for many readers, familiar in its plot if new in its disciplinary setting. Across many academic disciplines, scholars have become newly interested in materiality—what things are, how things are embedded in their environments, and how material qualities of things focus our attention in certain directions. The contrast between physical and digital things has inspired some of this recent attention. In literary and textual studies, for instance, scholars had long tended to locate a text's thingness in its content—the words—and not in its material instantiation as a printed book. But working with digital editions has prompted wider scrutiny of material qualities across expressive modalities: digital things, print things, audio things.[30] As another example, in human-computer interaction, comparisons between physical and digital have extended into investigations of purely digital materiality.[31] Concurrently, perspectives that decenter the human—that disrupt habitual understandings of thingness—have filtered outward from sociology, anthropology, and philosophy.[32] Throughout the time period condensed into this essay, these intellectual currents were strengthening. Just as I began to see thingness everywhere in my own academic domain, so, it seemed, did everyone else. I have participated in some of these conversations, and I have definitely learned from them. I have also observed, across multiple disciplines, a similar sense of awakening. Literature isn't just words! Digital things are also things! Nonhumans have agency! Human practices enact divergent ontologies![33] Of course they do! How could we not have seen this before? Thank goodness, we see more clearly now.

I have a name for this phenomenon. In the literature of knowledge organization, design rationale is called *warrant*. There are many kinds of warrant, typically associated with source of the rationale: the literature being organized, the state of scientific consensus, the users of an information system, ethics.[34] But there is a hidden warrant that is used more often than any of these. I call it the *duh* warrant. The duh warrant encompasses all the decisions that we don't think about at all, the ones that don't even seem like decisions. The duh warrant is everywhere. Fiction and nonfiction kept apart? Duh! What else would we do? Clothing in department stores separated by gender? Duh! There's no other way to do it, is there? Past decisions made by duh warrant tend to be the ones that seem exceptionally silly—or immoral—to us now. Well, *some* of the decisions made by duh warrant fall into that category. Many more decisions made by duh warrant continue to operate, quite without our notice.

There is a thrill to calling out the duh warrant. Your eyes have been opened! It feels like a big accomplishment. And it is . . . but the reality of the duh warrant is that it, like the mysteries of thingness, can only be dealt with provisionally. The duh warrant is a warning: it shows how the clear and obvious of today becomes the folly of tomorrow. It may seem pedantic, in the current moment, to articulate the justification for what seems to be a *duh* position. Indeed, it's harder than it seems to recognize that a duh position is a position at all and that it could ever be otherwise. But every time that we congratulate ourselves for identifying and reconsidering the duh warrant in the past, we'll want to take a good look at our present also. Duh?

4
INTEROPERABILITY

This chapter explores the related ideas of *reproducibility* and *interoperability*. In science, reproducible procedures are intended to enable the collection of confirmatory data toward the acceptance or rejection of a hypothesis. Reproducible data collection protocols enable interoperable datasets, facilitating the aggregation of data from multiple sources and the reuse of data across contexts. In concept, reproducibility and interoperability seem like straightforward and reasonable goals. In practice, their implementation has been elusive.

In the adventure essay, I tell stories about

- Cooking rice.
- French bread in different languages.
- The provenance of an afternoon pastry.
- A Thai robot curry taster.
- Protocols for removing dangerous posts from social media.

In the reflection essay, I ruminate over

- Scientific reproducibility as an unsettled concept.
- The challenge of semantic data interoperability.
- Empirical outcomes in the pursuit of interoperable data.
- The imperative of data criticism.

ADVENTURE: I CAN'T COOK RICE IN COPENHAGEN

Cooking Rice

It should have been a lovely meal, one day in late September. The cauliflower sabzi smelled divine. But the rice that accompanied it had disintegrated into a sodden mass. I poked a fork into the starchy lump on my plate, grimacing in frustration. We had been in Denmark four months already, and somehow I still had not managed to produce a decent batch of basmati.

It's true that rice can be fiddly, and burners always vary: for every stove, one has to find the setting that will keep a gentle simmer as the grains steam. But I had anticipated minor adjustments, not wholesale failures. Even as I had mostly adapted to the unresponsive radiant stove and the thin-walled pots, the rice flummoxed me. One day it would be mush; the next, hard and crunchy.

It was galling. So I worked at it. Methodically, I attempted to isolate the problem, carefully considering all the typical variables. I kept a close watch on the temperature. I reduced the amount of water; I increased the amount of water. I washed the rice thoroughly until the starch disappeared; I didn't wash the rice at all. I used different varieties of rice and different brands of the same variety. The problems persisted. Was the extremely hard water in Copenhagen affecting absorption? Was the heat of the burners too inconsistent? Finally I resorted to desperate measures. I bought a multifunction pressure cooker on sale from Amazon.de. It wasn't just for the rice, I rationalized. The pressure cooker would also be handy for soups and stews in the long winter months. Or so I told myself.

When I finally produced acceptable jasmine rice using my new appliance, I thought that I had solved the problem. Probably the stove had been the culprit all along, I reasoned. But with the cauliflower sabzi, I had tried basmati rice in the pressure cooker, and it had practically turned into porridge when I attempted to fluff it. Gah! These inconsistent outcomes were embarrassing, as well as vexing. It was, after all, a simple, familiar process, conducted under similar circumstances with similar materials. I wasn't coping with extreme conditions, like an open fire in the wilderness, or even with some unfamiliar rice variety. I was cooking in a renovated apartment in a European city with globally available ingredients! Why was it so difficult to isolate the cause of my annoyingly variable results?

Over time, as I mentally cataloged and compared my attempts, I began to realize that my process for cooking rice was actually much less rote and much less standardized than I had previously appreciated. All the while, I had thought that I was following a simple, consistent process: that I did the same thing every time, and that what I did was what everyone did. Bring water, rice, and salt to a simmer. Cover, turn down the heat to low, and steam for 20 minutes. Turn off the heat and let stand for 5–10 minutes. That's it!

But when I decomposed each step, I discovered an array of small adjustments that I might be making, usually without thinking about it. For instance, when I was taking the rice to its initial simmer, I would usually use medium-high heat . . . but why? Was that part of the "standard" process? When recipes do provide guidance here, instructions vary. Indeed, as I looked around on cooking sites online, I realized that lots of people, just like me, thought that they did the same thing every time. But processes were surprisingly varied (different ratios of water, different steaming times, different cooking vessels), and everyone articulated certain details while leaving out others (some people carefully described the size of pot per volume of rice, for instance, while others did not consider this worth mentioning).

In fact, the more that I investigated, the less settled and standardized the process became. There were so many factors and so many ways to combine them. My whole understanding of the situation began to break down. At first, I had imagined myself to be in a simple debugging situation: find the problem and eliminate it. Gradually, however, I came to believe that there was no error to fix. Instead, the rice-cooking situation in my apartment in Copenhagen involved many subtle deviations from my accustomed environment, and these variations combined in ways that remained difficult to understand and even to articulate. Moreover, where I had hoped that introducing an automated device into the process (the pressure cooker) would clarify the situation, the device had contributed its own set of characteristics into the mix, which were also unfamiliar to me. In some ways the pressure cooker made things worse, even as the rice improved somewhat, because I didn't have as much insight into what it was doing and how to manipulate it. Eventually, we did end up eating acceptable rice in Copenhagen on a consistent basis. But the texture of our basmati, in particular, was never as elegantly, toothsomely fluffy as it had been at home. Or so it seemed to me.

Because I found this lack of resolution deeply unsatisfying, I kept trying to isolate a single source for my difficulties, ignoring what I should have understood much earlier. Any home cook knows that adaptations always need to be made. The size of an average onion is relative; a stainless skillet takes heat differently from a cast iron skillet; chilies vary in their punch; and so on. Nonetheless we regularly mistake the idiosyncrasies of our personal environments as more typical than they really are. Those of us who imagine ourselves knowledgeable are perhaps the most likely to be obtuse in this way. In my case, I was expecting to make allowances for a different stove. But I was oblivious to other factors I'd been taking for granted: equipment, materials, procedural steps, even the characteristics of a successful product.[1]

In short, I simply hadn't noticed the extent to which my understanding of "basmati rice and how to make it" was really just "the way that I personally make basmati rice in the environment that I am used to." What I had imagined to be universal tacit knowledge of standard procedure was really just my ingrained habits. In my accustomed environment, my habits had produced reliable results. But in a new environment, the unstable foundation of this reliability revealed itself. Of course "the same process" of cooking rice was producing different outcomes in Copenhagen. It had *never been* the standardized process that I had imagined it to be.

But couldn't my new pressure cooker offer new possibilities? I imagined an alternate rice-cooking future, one where I would control everything. I would keep the same equipment and materials everywhere I went—for example, by taking the pressure cooker with me always and by always using the same brand of rice, along with filtered water and a specified purity of salt. I would perform a series of blind taste tests to determine the procedure that led to the best results, and I would do the same thing, rigorously, every time, in my completely controlled setup. I wouldn't just *think* that my rice-making process was standardized; it really *would* be standardized. In this future, I would have the exactly the same rice every time, no matter where or when I was. The same! Always!

I was taken with this idea for a moment. But then I remembered the rice of my childhood: Uncle Ben's Converted Rice. When I was growing up in the 1970s in southern California, Uncle Ben's *was* rice. I knew nothing else. Converted rice has been processed and parboiled; when cooked, grains of converted rice remain separate. Until I was about nine years old, this was the only texture that I thought rice could have. Then, in the fourth grade, I had dinner at my friend Carli's house. Carli was Korean, and the rice at Carli's house was totally different. It was sticky and soft and yet somehow firm at the same time: so delicious. At first, *rice* and *the rice at Carli's house* were separate concepts for me. But my perspective had begun to transform. Within a few years, what had been *the rice at Carli's house* became merely *rice*, and I had practically forgotten that Uncle Ben's existed.

During each of these stages in my life, *rice* did not seem like an especially fluid concept. When everyone I knew was eating converted rice in the 1970s, we did not imagine that our habits were unusual or that our world was small. In 1979, if a precocious me had conducted blind taste tests to develop a standardized rice process, I would not have thought that I was merely describing the preferences of a certain slice of middle-class, White America. Similarly, by the time I graduated from high school, I would have thought that, where rice was concerned, I *had* been close-minded but *now* I was enlightened. Real rice was (east) Asian rice!

How many times have I undergone a similar evolution, each time thinking that I had finally understood the true reality? Countless times. Even in Copenhagen of all places, I had multiple epiphanies regarding the rice universe. In addition to the difficulties I had cooking it, I had been mystified at not being able to find "basic" Asian rice, as opposed to jasmine or basmati, which I considered distinct subvarieties. Where is the Calrose rice, I wondered. And then it hit me: what I had considered basic Asian rice is all grown in California (the Calrose variety) and it is really the *American* version of basic Asian rice. People in Asia don't eat Nishiki or Kohukuo Rose brand rice; Asians in America eat Nishiki and Kohukuo Rose. Oh. Of course.

My brief fantasy of a real standardized rice process now seemed kind of silly. Sure, I would have the same rice every time, and I would spare myself the potential of disruption and disappointment. But the standardized process would also lock me into limitations that I wouldn't even realize were there. Whatever assumptions I had at the moment—about rice varieties, desirable characteristics, brands, and so on—would be with me forever. It would be like staying in my small world of 1970s Uncle Ben's Converted Rice for no reason except complacency.[2] How unfortunate that would be.

French Bread in Different Languages

It was in this frame of mind that I found myself drawn into a discussion about French bread in my Danish language class. For English speakers, Danish is an easy language to read and write, but its pronunciation is mystifying. One evening in class, we were attempting to pronounce *rugbrød* (the Danish for *rye bread*, which is a very common word). None of us could do it, and we were all laughing about how silly we sounded. Of course, then we were all thinking about bread, and so, during our class break time, one of the other students, Maria (from Venezuela), was asking the rest of us if we liked the bread in Denmark. Another student, Lucy (from Korea), replied vehemently: she hated it. I was a bit curious about this reaction, so I asked Lucy if she liked any bread at all. The bread in Denmark was just too *dense*, she replied, her lip curling with distaste. Lucy hesitated for a minute, and then she added that she did like French bread, but she gave me sort of a wary look, as if she didn't think I would understand what she meant.

At first, when Lucy used the term *French bread*, I thought she meant baguettes like they make them in France: light and airy but with a crisp crust. But when Lucy gave me that strange look, I had a sudden realization. I knew with certainty that when Lucy said *French bread*, what she meant was a soft, puffy bread that you would never find in France. So I told Lucy that I knew the kind of bread she meant, because when I had lived in Austin, Texas, I had lived near a bakery named *Tous Les Jours*. And Lucy said "Tous Les Jours! Yes!" And then we smiled at each other.

Despite its pillowy bread and pastries, Tous Les Jours described itself as "an authentic French bakery." This was confounding to many Austinites. But Tous Les Jours was indeed extremely authentic—for a *Korean* French bakery. It was a chain direct from Seoul, which was how Lucy knew it.

In the brief exchange between me and Lucy, we have one label, *French bread*, and multiple concepts: French bread like in France and French bread like in Korea. But of course there are many more concepts of French bread than that. After class, I thought about that wary look that Lucy had given me and how she had seemed to expect that I wouldn't understand her. I wondered what *French bread* might mean to Danes. My Danish dictionary defined *franskbrød* as "bread made with white flour, yeast, water or milk, salt, and fat." I learned something from that definition—baguettes in France do not include milk or butter, so *franskbrød* in Denmark is probably not what you would find at a boulangerie in Paris. However, those ingredients—white flour, water or milk, salt, fat—can produce breads with different textures, depending on proportion and technique. Intrigued, I searched Google Images for *franskbrød*. The results looked similar to American *white bread:* a soft loaf made with white flour and usually a little milk and butter in the dough, baked in a pan (see figure 4.1).

This constellation of concepts and labels had quickly become complicated! We had the following:

- French bread in France (baguettes)
- French bread in Korea
- French bread in Denmark
- White bread in the United States

We're probably familiar with these kinds of situations, no matter how many languages we speak or how many countries we've lived in. We might

Interoperability

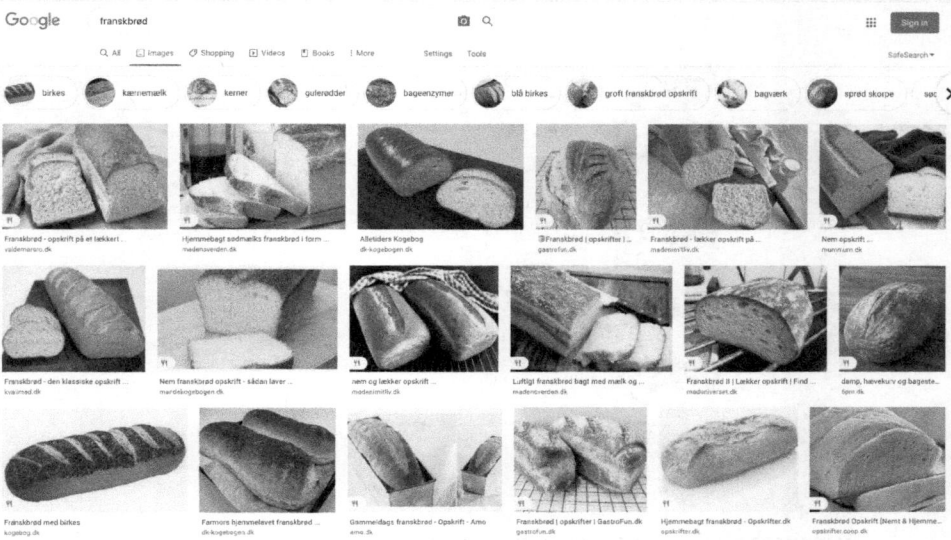

Figure 4.1
As shown in the results of an image search, Danish *franskbrød* looks a lot like American *white bread*.

have a single label—French bread—used to designate a variety of different concepts, or we might have multiple labels—white bread, milk bread, franskbrød—that designate a similar concept. (The former is called *polysemy* and the latter is *synonymy*.) In everyday life, we're usually happy to coexist with indeterminacy here. We assign some kind of meaning to *French bread* when we encounter it, assume that it's reasonably correct, and move on. Sometimes, as in my conversation with Lucy, we need to stop and negotiate a shared agreement on what we mean. But we typically do not need to interrogate the possible senses of every word or concept that we employ. We live with tremendous levels of uncertainty all the time, and we usually accept it.

But is it different when we are expecting that information systems are providing reliable information? What if we were in Austin, and we searched Google for a French bakery and then found ourselves walking into Tous Les Jours (see figure 4.2)?

Maybe we would be surprised. Maybe we would be delighted or intrigued: a happy surprise. Or maybe we would be confused, or annoyed, or both. We might accuse Google of getting it all wrong and wasting our time. We'd complain, perhaps, that Google should at least tell us the difference between

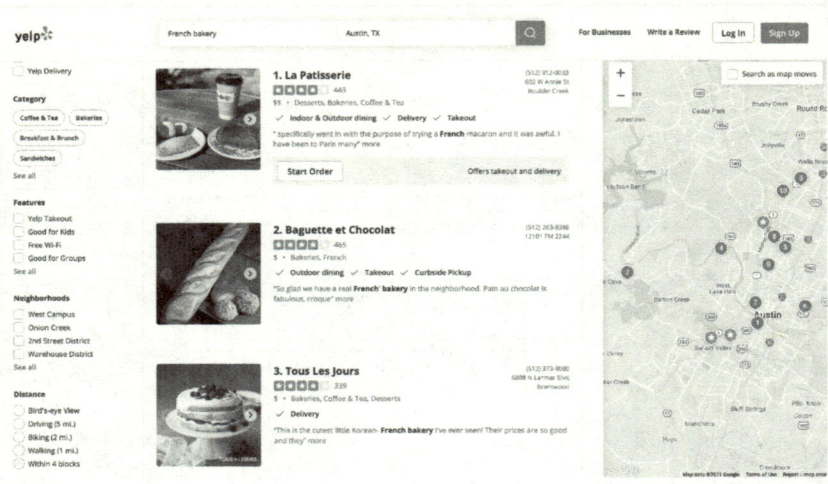

Figure 4.2
A Yelp search for *French bakery* in Austin includes Tous Les Jours in the first three results. (A comment clarifies that Tous Les Jours is a Korean-French establishment, but the text is smaller and easy to miss.)

French (French) and French (Korean), so that we could have thought about it before we went all the way there and didn't find the baguettes that we were looking for.

So retrieval with a search engine is one task where we might want to systematically delineate different kinds of French bakeries. There are many such tasks where the validity of our success relies on assumptions about the kind of data we are using: that all the data defines a French bakery in the same way, for instance. What if we wanted to count bakeries of different types—French, Asian, and so on—in Austin or any other location, to track economic growth or for any number of reasons relating to research, policy, or decision making? Would we count Tous Les Jours as French, as Korean, as Asian, or as all of these? And would we encounter problems if we wanted to aggregate counts performed by various people in various places where different decisions had been made? (In Seoul, for example, maybe Tous Les Jours would be counted as clearly French, whereas an outpost of a French bakery from Austin, Texas in Seoul would be counted as American—not French at all.)

There are various technical devices for managing such problems: for systematically distinguishing and relating the concepts that we use to create data. A controlled vocabulary, for instance, defines concepts, associates concepts with labels, and shows relationships between similar concepts. If we map these relationships with careful precision, we can distinguish between many distinct variations of closely related concepts. For example, we might show our constellation of French bread concepts like this:

```
<Breads based on white flour>
   <Dense crumb>
       French (Danish) bread
       White (American) bread
   <Airy crumb>
     <Crisp crust and airy crumb>
       French (French) bread (baguette)
     <Soft crust and airy crumb>
       French (Korean) bread
```

Isn't that marvelous? In normal speech, we would confusingly refer to all these distinct concepts with the same label (French bread), but in the artificial language of a controlled vocabulary we can identify the different concepts, assign them each a unique name, and clarify how they are different, so that we know that French (Danish) bread and white (American) bread are similar and not very much like a baguette. Our goal here is rather like my rice fantasy in which I always use the same pressure cooker and ingredients. The controlled vocabulary is a technical device that helps us to ensure a replicable process of data creation, just as the pressure cooker is a technical device for ensuring a replicable process of rice cookery. In my rice fantasy, the pressure cooker helps to ensure consistent basmati; for data creation, a controlled vocabulary helps to ensure *interoperable* data, that is, data that has the same meaning and structure from one dataset to another. With a technical device like my bread vocabulary we can always be certain about what kind of bread we have, whether we're trying to find some particular type, counting all the bread that exists, or whatever you fancy.

The Provenance of an Afternoon Pastry

That's the idea anyway. In practice the fantasy proves impossible, even as it seems so simple. The world keeps enlarging itself, subverting one's plans, wrecking one's lovely vocabulary maps. For instance, when my energy was flagging one afternoon in Copenhagen, I went for a stroll down to Islands Brygge and ended up buying a pastry at a bakery, Andersen. When I was passing by, I didn't know anything about Andersen, and at the time it seemed like a normal Danish bakery (see figure 4.3).

Figure 4.3
Andersen Bageri in Copenhagen.

A little bit later, as I was savoring my spandauer with a cup of tea in my office, I looked up Andersen on the Web to see what other people had thought of it. Had I selected a representative item? Do they have a specialty? I discovered that Andersen had won a contest for the best cake in Copenhagen 2019. But I also discovered that Andersen Bageri is not quite a normal Danish bakery. Andersen is owned by the European subsidiary of a Japanese bakery conglomerate, Andersen Group. Andersen Group holds a number of Danish-inspired bakery chains, mostly in Asia but also in the United States. These subsidiaries interpret the Danish bakery for the Asian market, just as Tous Les Jours interprets the French bakery for Asian tastes. But the Andersen Bageri in Copenhagen is different. To the casual visitor, it doesn't seem Japanese at all.

If we put something like Andersen in our constellation of French bread concepts, it would be as though the Korean owners of Tous Les Jours opened a bakery in Paris called Flaubert, where they employed French bakers to produce baguettes and croissants just like everywhere else in France. Would that Parisian bakery just make French (French) bread then? Or would its origins make it Korean (French)? Or would it be more precise to call it French (Korean) (French)?

Making distinctions is the whole point of describing things (that is, creating data about them). But making distinctions as fine as French (Korean) French or Danish (Japanese) (Danish) gives the whole enterprise an air of ridiculousness. Can't I go on an afternoon walk and get a lovely pastry without breaking down the whole system?

A Thai Robot Curry Taster

I can't even get some Thai take-out without breaking down the system. In Copenhagen, we lived in Amager, which many consider to be a boring neighborhood. Even in Amager, there were at least five options for Thai take-away within the immediate vicinity of our apartment. We went to one of them on the first evening that we arrived in the country (see figure 4.4). It was one of the few restaurants that I've been to in Denmark where the people working there (two older Thai ladies) didn't speak English. But Thai food is Thai food, right? We managed to get some pad kee mao and green curry that night without too much trouble.

Our Thai take-out in Copenhagen tasted familiar to me. It may be more difficult to source Thai basil in Denmark—the Thai restaurants tended to

Figure 4.4
The Thai restaurant on Østrigsgade where we got our first meal in Copenhagen.

use fewer fresh herbs—but otherwise it was similar to what we might get in Durham or anywhere else in the United States. And the women in the kitchen at the Tuk Tuk were definitely Thai. So it seemed like reasonably authentic Thai food. Not Thai (Danish) or anything like that.

But have I ever been to Thailand? Well, no. And apparently, for Thais, pretty much all the Thai food outside of Thailand is unpalatably sweet. In fact, when Yingluck Shinawatra was prime minister of Thailand, she found the situation with Thai food around the world to be so frustrating—if you were outside of Thailand, you just didn't know what you might be served—that she started a government initiative to establish authenticity standards for Thai dishes (Fuller 2014).[3]

Such an authenticity standard would aim to

1. Unambiguously define a concept (such as the authentic version of a particular noodle dish).
2. Label that concept (*authentic* pad kee mao or Thai [Thai] pad kee mao).
3. Distinguish that concept from others (by, for instance, defining a separate concept of *inauthentic* or Thai [foreign] pad kee mao).
4. Show how concepts are related (establishing that the inauthentic, foreign pad kee mao is sweeter than the authentic version).

Once concepts of authentic pad kee mao or tom kha gai have been clearly defined and distinguished from their inauthentic counterparts, one simply collects samples from Thai restaurants and compares each sample to the reference standard for the authentic dish. Samples that are sufficiently close to the standard can then be reliably pronounced as authentic. The results of such assessments might inform overall restaurant authenticity grades, which could then be publicly posted. Thais around the world might then confidently select a place to dine. Such was the goal of the Thai Delicious committee, which was formed to carry out Yingluck Shinawatra's initiative.

Conceptually, this project is similar to mapping the different concepts associated with French bread. We're just making a kind of dictionary: a more careful and systematic dictionary that seeks to account for and eliminate the ambiguities of natural language. But with the French bread, I felt, at least initially, that the universe of concepts was easy to comprehend and explain. It just seemed straightforward and obvious. When I considered the different concepts associated with French bread—Korean, Danish, French—they almost arranged themselves into my taxonomic map according to their own latent structure. Or that's what it felt like to me. All I needed to do was look closely with careful attention and the permutations of crumb and crust simply revealed themselves. Likewise, it was easy to ignore pesky implementation details like how to decide whether a particular loaf would fall into one class or another. Looking at my orderly vocabulary map, it was hard to imagine that loaves with an airy crumb (Korean French bread) wouldn't be obviously distinct from loaves with a dense crumb (Danish French bread or American white bread). In the map, those concepts were clearly delineated, and they seemed self-evidently different.

In the initial flush of creating my lovely vocabulary map, however, I didn't consider that I might have deceived myself. Couldn't it just be possible that my map was obvious to me . . . but not to everyone else? That thinking that I had simply revealed what was already latent within those concepts was just a weird way of both amplifying and disavowing my own skill at the same time? On the one hand, I made it seem as if that taxonomic map of French bread was something that I discovered, like a scientist, through focused observation. The world was cleaving itself into its own taxonomic joints, and I was its mere recorder. On the other hand, this map that I have created is so obviously correct that I don't need to explain or justify it. It's not a set of choices that I've made or preferences that *I* have. It's just the concepts themselves, as insightfully perceived by a taxonomist of skill and flair! Of course, I didn't consciously think this when I was doing it. It was all just easy and effortless. I created, disambiguated, and related the French bread concepts in exactly the same way that I cooked rice before coming to Copenhagen: not realizing how much of my accumulated experience and skill was rolled up into what seemed like the rote performance of a universally understood task.

But with the Thai food, someone else was creating the system according to their own self-evident understanding, and—surprise! it seemed much less straightforward. I didn't automatically understand the division between authentic and inauthentic—yes, I know that pad thai often has a little sweetness, but is *all* the Thai food that I've ever eaten *that* disgustingly sweet? Certainly the Thai cuisine that I've experienced has varied a lot. Sure, I've eaten my share of cheap and greasy Thai noodles in hotel rooms (my secret indulgence when I attend academic conferences). But not all of it has been like that! Am I to understand, nonetheless, that all the Thai food I've experienced is equally inauthentic, even as it seems to span a range of styles and levels of quality?

I can see that, to the Thai culinary experts designing the standard, the concepts of authenticity and inauthenticity seem clear and obvious in a way that I can't grasp. And I accept that a group of Thai gastronomes have more knowledge and standing in this area than I do. After all, when I was the person cleaving the world into taxonomic joints, the divisions between French bread concepts seemed clear and obvious. That clarity, however, was achieved only through a kind of ignorance, in which I arranged the world in a way that seemed right to me and then assumed that it was the world's doing rather than mine. When I'm not the person making the cuts, and when my perspective is not taken as given, the sharply delineated distinctions that others may see so clearly don't seem so natural and obvious. Where the Thai experts might see obvious distinctions, I don't.

If everything that I know is lumped together as "inauthentic," and yet what I know encompasses a diversity of variation that I can nonetheless distinguish along various dimensions, then could it be that the boundaries between authentic and inauthentic are not as easy to describe as all that? It makes me wonder, as I didn't wonder for the system that I had myself designed, how this authenticity standard would be applied and who would be able to apply it. Clearly my palate couldn't manage that job. But whose palate would be up to the task in the way that the designers might have idealistically envisioned? Indeed, if palates weren't variable and tongues fickle, this discrepancy between Thai food in Thailand and elsewhere would never have arisen, and it wouldn't have been necessary to design a standard. Parallels with my rice-cooking travails seem pertinent. Not being able to replicate "proper basmati" in a new place made me realize two things: first, that the idea of "proper basmati" in my mind had been rather exquisitely calibrated to the circumstances of my accustomed environment in ways that I had not fathomed, and second, that my supposedly simple, standardized process for making "proper basmati" was likewise deeply integrated with my personal habits to an extent that was, again, previously unrecognized by me. Wouldn't the same thing happen to a corps of Thai food tasters around the world? Maybe, as I had done with the rice, they would think that they all knew what authentic Thai food was supposed to be. But really, they would all understand it slightly differently, on the basis of their own interpretations of the standard. Then, as they went out into the world tasting and assessing, the American or Danish or Chilean versions of Thai cookery that they ate would subtly affect their perceptions and their judgments would start to shift . . . perhaps very slightly. But the authenticity assessments would gradually show the effects of the tasters' changing circumstances and resulting interpretive adjustments. The authenticity assessors would believe that they were replicating a standard process to collect interoperable data, but they would actually be making their own interpretations according to local circumstances—not quite the same thing.

Ah, but remember how I fantasized about creating a location-independent bubble, partly through the standardization possibilities enabled through my pressure cooker? The Thai Delicious committee had a similar vision. They would automate the implementation of authenticity assessment by using robots! A human being's palate changes with every taste, but a robot taster's palate is constant. Moreover, a robot can replicate its processes with much greater fidelity than even the most careful human. But where I merely imagined a situation in which I transported my pressure cooker and all ingredients from place to place, thus sparing myself the indignity of bad rice, the Thai Delicious committee actually created a working prototype of a robot taster to assess the authenticity of Thai green curry (Fuller 2014). First, the committee devised a reference standard for authentic Thai green curry, based on expert opinion. Then, they mapped the standard to a range of chemical characteristics that could be tested by machine analysis. The robot taster compared a test sample to the reference standard and assigned the test sample a similarity score in the range of 1–100. If the test sample scored less than 80, it would be designated *not authentic*. The *New York Times* Web site, where I read about this project, posted a short video that showed the robot taster at work. The *Times* reporter had the taster assess the green curry from the Thai foreign correspondents' club in Bangkok. That curry received a score of 78, so it failed the test.

When I read the story of the Thai robot curry taster and saw the video of the green curry from Bangkok being judged as not authentic, all my thoughts about the rice and the French bread and what we hope to achieve in these kinds of situations began to coalesce. The marvelous thing about the robot curry taster is that it is completely reasonable *and* utterly absurd at the same time. Of course the robot curry taster is rubbish! Designing a robot to analyze the chemical composition of a dish in comparison to a reference standard completely and fundamentally misunderstands what people are doing when they taste a curry, look at a picture, hear a song, use language, or make sense of human concepts through any systematic symbolic means (i.e., interpret data in any form). Using the robot curry taster to establish authenticity is akin to assessing the quality of a sonnet by comparing its actual rhyme and meter to the standard sonnet pattern. Your reference standard in both cases is selective and limited: only partially expressive of the whole. The taste of a curry results from the synthesis and integration of its parts, not from the percentages of each component. The experience of a poem emerges from how the constraints of rhyme and meter entwine with all the other choices of style and sense that the author has made. Moreover, a superlative poem, or a really great curry, challenges its pattern, adhering to it in some ways but not in others. Expressions push against their genre conventions in unpredictable ways, though. Sometimes they make lots of tiny deviations in many dimensions, and sometimes they make a single significant change. Too, the variability of the interpreter is actually a key element of how we make sense of any kind of expression. The way that humans comprehend notions like the *Thai-ness* of green curry is dynamic and contentious, and that diversity of response is important; that's how traditions evolve and cultures shift. It is, in other words, a *feature* of human capacities that our understanding of green curries and sonnets is not rigidly consistent, and that the same person will taste the same curry differently over multiple iterations. That's how our cultural traditions remain vibrant and alive.

It would be easy to continue in this vein, because the robot curry taster is so hopelessly misguided on multiple fronts. What an awful idea! And yet . . . the robot curry taster is, at the same time, perfectly reasonable. That's what I love about it. Will the robot curry taster ensure greater cohesion and consistency in the curries that pass the test? Yes! Yes it will, and that's not meaningless. Will some delicious curries fail the test? Of course. These failures will seem especially egregious when they involve novel dishes created by masters of Thai cuisine, who know perfectly well that they are deviating from a generic, homogenized view of "authenticity." It is also likely that the taster's results will seem irrational when curries that seem very similar to those that pass are deemed to narrowly miss the mark, as happened with the curry from the foreign correspondents' club in Bangkok that the *New York Times* reporter submitted for a test. So the robot taster's results will be just as contestable as human judgments but with a different logic: the robot will seem wrong because it replicates its process too faithfully and is not sensitive to deviations that have coherent and understandable motivations (such as the adventurous palates of skilled and knowledgeable cooks, among other singularities of local context). Moreover, if the reference standard is not regularly recalibrated, will the results seem more even more arbitrary over time, as tastes change? Yes, the assessments of the robot taster will appear to degrade over time, as culture changes around it. All this is true. But will a frustrated Thai person who uses the ratings on foreign vacations be less annoyed by overly sweet curries? Undoubtedly yes! Does this mean that curries that pass the test are actually more authentic than those that don't? Certainly not. But within a narrow scope, ensuring a more limited range of variation between instances does have utility.

Controlled vocabularies and other technical devices that aim to facilitate the replicability and interoperability of data are just like the robot curry taster. The map of French bread concepts that I made has lots of problems: it's reductive, it's not comprehensive, and it reflects a single perspective that some people might disagree with, mildly or strenuously. It's also limited in the amount of control that it can achieve, because people will understand and apply it differently in practice. Likewise, however, my tiny controlled vocabulary does separate concepts with more precision than we typically do, and it explains the characteristics that it uses (crumb and crust) to establish concept definitions and relations. Within its purview, will such a controlled vocabulary produce data that is more internally consistent and interoperable across contexts? Yes, it will.

Depending on the situation, the finite level of control that we can achieve from technical devices like this can be the difference between data that we'd be fools to trust for even silly, mundane purposes (like finding French bakeries in Austin) and data that we might rely on to make serious decisions (like how to enforce antidiscrimination laws) or to tell us something real about the world (like determining whether terrorism is increasing).[4] Fundamentally, although the technical devices that we employ to ensure interoperable data are inherently flawed and do not work in the way that we might hope, they are nonetheless the difference between ambiguous, uncertain, terrible data and slightly less ambiguous, uncertain, not-quite-as-terrible data.

Protocols for Removing Dangerous Posts from Social Media

Not-quite-as-terrible data is, I acknowledge, an uninspiring goal. Especially when we are making significant decisions or hope to provide solid evidence of something—when results really seem to matter—it seems like we should be able to aim higher. For instance, social media platforms such as Facebook and Twitter have struggled to identify postings that foment acts of violence, to develop policies for removing such posts, and to devise sufficiently replicable procedures (either manual or algorithmic or some combination) for implementing those policies. Identifying dangerous social media posts on global platforms is conceptually very similar to identifying authentic Thai food. The consequences of inaction, however, are dramatically different. If a Thai person traveling abroad is unsatisfied with dinner, nothing else happens. But when conditions are ripe, dangerous posts on social media can lead to catastrophic violence.[5] Accordingly, it's easy to acknowledge the limitations of something like the robot curry taster. So what if its pronouncements are fundamentally invalid? No one will suffer much from a meal they don't like. In contrast, because dangerous posts may have much higher consequences, it seems unhelpfully pessimistic to pre-emptively assert that our options for managing such situations—technical devices to constrain interpretation—are intrinsically limited in what they can achieve. The point, however, is not that such technical devices are worthless; they are both useful and necessary. But the astounding diversity of the world will always break our technical devices, and when that happens, doing one's level best to follow the supposedly replicable process will only result in porridgy rice—accompanied by frustration, embarrassment, and weird guilt. No one will be happy, and justifying one's actions by saying that you followed the rules won't help. (Believe me, I know.)

In other words, a dilemma like sorting social media posts into acceptable ones and dangerous ones can never remove the need for human judgment in light of local circumstances. Moreover, characterizing the problem of identifying dangerous posts as merely replicating a universal procedure—analogously to transporting our rice-cooking setup with us everywhere we happened to go—is unlikely to produce the outcomes we hope for.[6] It is true that developing specific adjustments for local conditions can improve the precision of a technical device, but special rules for particular circumstances can do only so much. The world contains too many situations like Andersen bakery in Copenhagen: those that align with no existing rule or with too many rules or between rules. These are the kinds of situations in which people need to control (and perhaps overrule) the technical devices rather than being controlled by them.

Facebook's approach to such situations is a classic one. Facebook, which relies on human moderators to assess dangerous posts, has developed elaborate rules to define what counts as *dangerous* in different localities and to ensure consistent application of these rules. Such rules are a form of technical device to ensure replicable data generation, often used in concert with controlled vocabularies. Library catalogers and other information professionals have been using and developing such processes for eons. But library catalogers know that the *taste and judgment* of the individual practitioner is a necessary element of the process and that replicability is therefore limited and relative. Indeed, all empirical studies of people creating data come to similar conclusions: such work is never a matter of rote replication. Just like people cooking rice, we're always making tiny, unarticulated adjustments that are exquisitely tuned to our environments. The often unacknowledged role of individual flair can sometimes cause us problems, like gloppy basmati in Copenhagen. But flair is vital to any enterprise of data creation. It's how we adapt along with our environments.[7] In contrast, when systems like the robot curry taster attempt to automate these judgments and remove the human aspect of interpretation, the actual rigidity of their judgments quickly starts to seem irrational.

If it still seems unsatisfying, that's because many of us would like knowledge to be stable, and if we accept that knowledge is often variable and conditional, we lose our idealism and our power. When I kept failing at the rice, I was, to be honest, humiliated. I felt unmasked as an impostor, someone who didn't actually know what she was doing. I wanted better rice, but I also wanted to retain my sense of mastery and control. Maintaining the fiction that I was executing a replicable process would enable me to keep that feeling of stability.

The desire for control is not in itself a shameful one. But it is hubris to imagine that control is stronger than its limits, and hubris blinds us to reality. In reality, we are constantly wrong, we constantly disagree about what to do next, and our rice is never as consistent as we thought in the first place. The technical devices that we put in place to help us along—the processes, the rules, the maps, whether automated or manual—are guides, not gods.[8] We should let ourselves be guided by our technical devices, assuredly . . . until we cannot be. When the "replicable" process delivers terrible rice (or, um, data), we shouldn't accept it, nor should we assume that the problem is in our faulty implementation. But "perfect rice every time" is a mirage also, even if we do manage to achieve the same result consistently. And if that seems weird to you, just imagine setting the robot curry taster loose on the restaurants of Bangkok after it has slept in a drawer for 25 years. Would you trust its results? If you would, unquestioningly, well . . . I have a wonderful pressure cooker to sell you. It makes perfect rice every time . . .

REFLECTION: HOW DOES THE SAME PROCESS LEAD TO DIFFERENT OUTCOMES?

Scientific Reproducibility as an Unsettled Concept

The *reproducibility* of an experimental procedure is a key component of modern scientific practice. The result of a single experiment is suggestive but inconclusive; only the aggregated results of multiple experiments provide sufficient evidence to confirm a hypothesis. Preferably, many scientists perform the same experiments independently. If results concur, reliability of the findings is established. If results differ, the initial findings are reassessed.

Although this account seems straightforward, the reality of scientific reproducibility is more complex. In theory, by documenting data collection and analysis protocols in published papers, scientists enable others to replicate experiments and verify results (Goodman et al. 2016). In practice, reproductions are seldom performed. Scholarly prestige, generally, rests upon new discoveries rather than the confirmation of reported findings (Tenopir et al. 2015). Researchers lack incentives to reproduce others' work.

Furthermore, the meaning of reproducibility in actual practice is underspecified and debated. Philosopher Nancy Cartwright (1991) differentiates between concepts of *replication* and *reproduction*, in which *replication* repeats an existing study as closely as possible in all dimensions—study environment, materials, procedures—whereas *reproduction* performs the existing study with some small variation of the existing setup. While the goal of replication is verification, the goal of reproduction is extension, to show how original results apply to a slightly different population, or in the context of a slightly different environment, or with another variation. Cartwright's proposal extends Henry Collins's distinction between *routine checking* (assurance that a study was performed correctly) and the provision of additional confirmatory results (Collins 1985).

In line with Cartwright and Collins, scholars generally agree that scientific reproducibility represents a spectrum of activities. However, opinions differ as to which actions have the most weight. Schmidt, for instance, suggests that strict replication of an existing experiment is of limited utility

(Schmidt 2009). Meanwhile, computationally intensive research has taken an opposite approach. In those fields, reproducible science has generally meant re-executing exactly the same code using the exactly same data (Peng 2011). Terminology, additionally, varies widely. Schmidt, for example, distinguishes between *exact replication* (Cartwright's *replication*) and *conceptual replication* (Cartwright's *reproduction*).

Essentially, it's never certain what anyone might mean when they discuss *reproducibility, replication*, or the like. In a study that I was involved with, for instance, we asked researchers from various disciplines to define *reproducibility* as a general concept (Feinberg et al. 2020). When our study participants considered reproducibility in an abstract sense, they tended to describe it as an exact replication: doing the same study again in the same way, to verify results. But later in our interviews, we asked participants about the benefits of reproducibility for them personally. When discussing reproducibility in this concrete sense, our participants tended to invoke a different concept: they emphasized reusing someone else's code or data for a new experiment or analysis. So reproducibility tended to mean different things depending on whether it was being discussed in an abstract, general way or in a specific, personal way.

In recent years, various disciplines have reexamined their practices regarding reproducible research, directing public attention toward this eminently fungible concept. As Collins (1985) emphasizes, no study can be exactly repeated, just as no river can ever be crossed in the same way twice. But some studies can accommodate greater levels of variation than others. All samples may differ, for instance, but samples of living organisms (e.g., people) exhibit greater variability than samples of inorganic matter (e.g., phosphorus). We can manipulate the phosphorus to achieve a certain level of purity, but we have no such equivalency standard for human beings, or snails, or radish plants. Similarly, research of living organisms that attempts to describe changes in mental or emotional states often relies on measures that are less direct than measures associated with physical states. We perceive it as less exact to measure the human experience of pain intensity and more exact to measure the purity of a phosphorus sample.

What it means, therefore, to replicate or reproduce a chemistry experiment may be quite different from what it means to replicate or reproduce a psychology experiment with human subjects. In 2015, such tensions led to a "reproducibility crisis" within the field of social psychology when a review of influential past studies contended that assertions of reproducibility had been overstated (Open Science Collaboration 2015). Some social psychologists endorsed this reassessment, seeing it as a corrective to shoddy research practices, while other interpretations described a power grab of sorts, in which previously accepted standards were questioned for the benefit of some and detriment of others (Dominus 2017). No matter how disciplinary conflicts like this are resolved, such episodes reveal significant differences in how reproducibility is understood and operationalized, even within a single scholarly community.

The Challenge of Semantic Data Interoperability

When reproducibility rather than replication is sought, researchers employ standardization—within disciplines, subject areas, and individual projects—to facilitate it. But as we might guess from the example of social psychology, empirical studies of scientific work practices generally demonstrate more variability and improvisation in standardized processes than practitioners realize. In particular, the recording of observations—that is, data collection—is often perceived by scientists to be more mechanical and require less knowledge and skill than the design of a study or the analysis of collected data. (Accordingly, as Macaluso and colleagues demonstrate in a bibliographic analysis, such tasks are more often delegated to women [Macaluso et al. 2016].) The anthropologists and sociologists who study science in action invariably establish, however, that data collection activities involve creative, interpretive work. Archeologists, for instance, determine the color of a soil sample by matching the earth that they are digging to the Munsell color chart, which displays a set of swatches of brownish hues. The chart is perforated with holes under each swatch, so that an archeologist in the field can put a soil sample on a trowel and then stick the sample under holes on the chart, to visually match the sample to the best swatch. This comparison of swatch to sample seems like a rote process. In Charles Goodwin's account, however, archeologists are employing myriad tiny decisions as they assess their soil samples (Goodwin 1994). The colors on the chart, as printed swatches, are generally shinier than the matte earth samples, and light is reflected from the swatches and samples differently. "If this soil were turned into a swatch, which swatch would it be?" is the real question that archeologists in the field are asking, and this question involves imagination, judgment, and the expertise of seeing how others on the team have answered similar questions for a particular dig.

Similarly, Bruno Latour (1999) recounted how botanists distill the forest into a cabinet of specimens; Geoffrey Bowker (2000) revealed differences in collection and testing of lake water samples over time; and Susan Leigh Star and James Griesemer (1989) noted competing concerns of trappers and naturalists collecting biodiversity specimens in the Sierra Nevada. What these and many similar studies show is that data collection is tuned to local environments, specific cultures, and individual style.

Notably, the practice of data collection remains creative and flexible even with the introduction of technical devices, such as controlled vocabularies, that are meant to standardize data expression and ensure semantic interoperability. Goodwin's color chart is one example: even as the chart limits the ways that archeologists can describe soil color, archeologists still need to interpret the sample in the context of the chart. Similarly, Kansa, Kansa, and Arbuckle (2012) have described how zooarchaeologists working in the same region in Turkey, using the same system to describe tooth wear on fossils, nonetheless structured tooth data so differently that skilled human editors could not easily aggregate it. Likewise, Jackson and Barbrow (2015) have discussed the insufficiency of an extremely detailed collection protocol for a single ecology lab to enable frictionless data collection. From a "big rock in the middle of the stream" to bears "disrupting the collection equipment" the world presents an environment that cannot be circumscribed by standards but must be interpreted in light of them (Jackson and Barbrow 2015, pp. 1773, 1772). Across such accounts, the same conclusion arises. Although data collection might be mundane and tedious, it is also dynamic and diverse. What's more, this situation obtains even when data is collected purely by mechanical means, without direct human intervention.

With the introduction of step counters, household sensors, and similar tools, automated data collection devices have entered daily life. In studying the uptake of these everyday tools, researchers observe similar behavior to that of scientists and other professional data collectors. People use fitness data trackers in a wide variety of flexible, creative ways, some of which involve keeping strict counts and others of which don't; people who live with household sensors may adjust habits to promote or avoid certain responses from their devices, and so on (Rooksby et al. 2014; Tolmie et al. 2016).

Nonetheless, in the context of data aggregation and reuse, a belief persists that scrupulous use of standards—reproducible procedures for collecting data, schemas for determining what kinds of data to collect, and vocabularies and units for expressing the collected data—will produce data that transcends the time and place of its collection. Data that adheres to standards in this way is said to be *interoperable;* it can be combined with other standards-compliant datasets without requiring significant preprocessing (or *cleaning* in the lingo of data science). Interoperability is often described as having multiple levels. A basic form of interoperability involves transfer protocols and file formats for data exchange and storage. Another form of interoperability involves structure and syntax of the expressed data; an example is using a standard format for expressing time and date. Semantic interoperability—data whose meaning is the same—is the most complex (Zeng and Chan 2010; Elings and Weibel 2007). Although structural interoperability might seem to facilitate semantic interoperability, this is not necessarily so. For instance, two datasets may use the same structure—such as a creation date for a photograph—and yet mean different things by that structure. In one case, a creation date might record the date and time when the photo was taken, while in another case, a creation date might reflect when a digital file was saved. Often, people who use data standards do not imagine that someone else might interpret the notion of something like a creation date in some other way. If data is generated as part of a digitization project for historical photographs, for instance, the creation date, in the context of that project, will only make sense as the time the digital file was saved. But if data is generated as part of a project to provide photographic evidence of an unfolding event, then the creation date will only make sense as the time when the image was initially captured.

Time constitutes a particular challenge to semantic interoperability. It can be difficult to track an evolving sense of underlying concepts. Michael Buckland describes how, although books don't themselves change over time, the subject headings assigned by library catalogers (phrases that indicate what the book is about) become obsolescent (Buckland 2012). Most obviously, this may occur when a term formerly viewed as value neutral becomes or is recognized as offensive. But subject headings evolve in many ways. When catalog entries were recorded on paper cards, for instance, space constraints of the card format enforced a limit on the number of assigned headings. When catalogs were digitized in the 1960s, this constraint was lifted. Initially, data collection practices for subject headings continued to reflect the older technical environment. Gradually, however, habits changed. Recently cataloged books—particularly fiction—may have ten or more subject headings. Older books, in contrast, have two or three. Does this change cataloging data? Yes, it does. A book with few headings typically employs more general terms (such as "US history") whereas a book with many headings includes more specific terms (such as "Japanese Americans—Evacuation and relocation, 1942–1945—Personal narratives" and "Concentration camps—California—Manzanar").

I find library catalogs particularly evocative because they are ordinary and ubiquitous and, at the same time, extremely high-quality, standardized data that has been successfully aggregated and reused for over one hundred years. It takes only a few minutes of poking around a large library catalog to establish that semantic interoperability is, in its ideal sense, an impossible dream. But most people have never noticed that the library catalog is, from a certain perspective, both excellent and terrible at the same time, and so maybe the ideal of semantic interoperability isn't a necessary one. (Likewise, is it really a big deal if I think I'm making the rice the same but I'm not? Typically, not.)

Of course, goals matter. Ayelet Shavit and James Griesemer (2011) describe similar challenges associated with a resurvey of Yosemite biodiversity data 100 years after its original collection by the Museum of Vertebrate Zoology (MVZ) at the University of California, Berkeley. Shavit and Griesemer explored how ideas of *space* and *location* were subtly different for the earlier and later researchers. Moreover, both of the human conceptions of space and location differ from the appropriation of space by the organisms that the humans were hoping to track. To be somewhat reductive, finding "the same place" where a population of rodents might have lived 100 years ago is much harder than finding a spot on a map (and finding a spot on a 100-year-old map is not trivial, either). But for the resurvey, precision is important. The scientists here are hoping to establish how biodiversity has changed over time, and if locations are not congruent, the data has less force. (Do we really know if the populations have increased or decreased, or have we just sampled the wrong spaces?) Here, an assumption about the interoperability of *location* might lead to poor decisions for conservation and land management.

Sometimes, as with my rice-cooking adventures, we are merely oblivious to the local adjustments that we make to standardized processes. Other times, we make local adjustments consciously, developing our own rules to specify how standards are implemented. Most situations lie in between. We know that we're adapting a standard for our own uses, and we may coordinate informally with our colleagues to do so, but we're not diligent about documenting what we're doing or thinking about consequences for future data use. This situation constitutes a tremendous challenge for reuse of scientific (or any) data: the work involved to make a local team's extemporaneous decisions legible to unknown future users is great, and there is little reward for it.[9] It's not impossible to reuse data that lacks explanation of its making, but this also requires extensive, often unappreciated labor.

Still, no matter how well documented and precisely fitted to standards, any data reuse will require some measure of this interpretive work. An insightful example in this regard involves an account of three later zooarchaeologists' independent analyses of an old, unpublished dataset from the 1960s (Atici et al. 2012). Each later zooarchaeologist interpreted the original data collector's identification of bone samples differently, which led to three different interpretations of what the data meant. All four zooarchaeologists—the three later analysts and the original data collector from many years ago—performed scientifically valid, correct work. And yet three different interpretations of the dataset exist, based on different ways of understanding what the data was describing in the first place. This kind of variability is all that we can expect from any data, even the "best" data that we might be able to envision.

Sigh. There is an old joke that all research in the social sciences can be summarized with the phrase "it depends." Meanwhile, the humanities accumulates ever more alternate readings of the same texts. This may be a realistic way of comprehending the world, but it can be wearying and feel powerless. It makes a lot of psychological sense that we would want to apply problem-solving, solution-oriented paradigms wherever we can.

In the field of information organization that is my academic specialty, classificationists used to approach their practice with just such a problem-solving orientation. Using terminology derived from mathematics, the influential classificationist S. R. Ranganathan, working in the early to middle twentieth century, would accordingly assert that catalogers who followed his "canons" and "principles" would have access to, for instance, "the true connotation of the term 'Medicine'." (Ranganathan 1959, 17). In other words, understanding "Medicine" is a problem that can be solved by following a set of rules. Today, pronouncements like Ranganathan's are dismissed as reductive and arrogant, and I agree with that assessment. But there is nonetheless an allure in being able to make such pronouncements, especially in the person of someone like Ranganathan, an idealist who wanted to make the world better by organizing it properly for once.

It is precisely this allure for certainty that has encouraged an increasing preference for automation as the means to enable reproducible science, particularly in the context of computationally intensive data analysis and modeling, which now occur across scientific disciplines (Kitzes, Turek, and Deniz 2017). In much of this data-intensive science, researchers are not collecting data via experiment or observation; they are performing computational analyses of existing datasets. To enable the checking of results, researchers have, therefore, been encouraged to make their code and data available to others. A replicator in this conception would rerun code (and not rewrite it). In consequence, practitioners of data-intensive science are urged to provide the exact code and data that they used to produce their results, rather than documenting a data-collection protocol, as has been more common in traditional scientific disciplines. Reproducibility is thus limited to an automated replication of analysis.

This preference for automation extends to data preparation (*cleaning*). This means that a task like standardizing date formats should, as a matter of best practice, be performed via an automated script rather than "by hand" (that is, by a person). Judgments about how to map one kind of format to another are then centralized within the script, which is theoretically sufficiently documented so that others can understand what it does. Additionally, any code required for either data preparation or data analysis should not just be provided and documented but, ideally, packaged together (for example, with container software such as the open-source Docker) so that it can be run without requiring the secondary user to recompile the code for their system environment.

There is nothing wrong with this approach, if we recognize what an exact replication like this enables: it means that someone else can redo your work exactly as you have performed it. We need to keep in mind, however, that rerunning a script on a dataset doesn't provide access to the interpretive processes that led the original researcher to choose one mapping over another, or make any other kind of decision. To return to the rice situation from the adventure essay in this chapter, my rationale for selecting a certain pressure cooker was opportunistic: it happened to be on sale. My "packaging" of my rice setup doesn't justify or explain the device, it just enables someone to repeat exactly what I did. There is definite utility in providing a black box that achieves the same results, of course, but without opening the black box it is not easy to extend the setup to other situations.

Empirical Outcomes in the Pursuit of Interoperable Data

If there's anything strange or surprising about any of this, it's that, despite the vast amount of empirical evidence regarding the limited nature of reproducibility, we are continually surprised by the limited nature of reproducibility. I am reminded of an incident from my first semester as a master's student in information science at the University of California, Berkeley. In our class on information organization and retrieval, we read a paper by George Furnas and colleagues about choosing names for menu commands and other elements of software interfaces. In 1987, when that paper was written, text interfaces (as opposed to GUI interfaces) were still common. In a text interface, users need to actually type (recall) the names of commands, rather than merely recognize commands when they see them in the interface. So human-computer interaction researchers tried to formulate best practices for command nomenclature. In the studies reported in the paper, researchers asked participants to name different sorts of things (some participants were asked to provide verbs to identify text-editing operations, others were asked for synonyms to label common objects, others provided keywords for information retrieval, and so on). Although the researchers expected some variation in naming, they also expected ultimate convergence on a *best* name. That convergence didn't happen. Ultimately, Furnas and his coauthors summed up their findings this way:

> Simply stated, the data tell us there is no one good access term for most objects. The idea of an "obvious," "self-evident," or "natural" term is a myth! Since even the best possible name is not very useful, it follows that there can exist no rules, guidelines or procedures for choosing a good name, in the sense of "accessible to the unfamiliar user." (Furnas et al. 1987, 967)

One of my classmates, Nathan Good, was skeptical about the Furnas paper. So he conducted an informal experiment. He asked other students hanging around in the basement lounge to label some common objects (e.g., a notebook and a plastic bottle of Coke). To Nathan's mild surprise, Furnas's findings were borne out. A variety of different names were contributed (e.g., notebook, folder, binder; Coke, bottle, soda). Of course, no one said that the Coke was actually orange juice. But people perceived the Coke at different levels of specificity (Coke, soda, drink) or focused on different aspects of its identity (for instance, the container (bottle) vs. the contents (Coke)). This was exactly what had happened in Furnas's paper. People did not merely come up with one set of synonyms for the same concept; that would actually be an easy problem to resolve. Instead, the names reflected different perceptions of what was being described and different aspects of the item. This is what made the problem devious. It's not more or less correct to name an item as "plastic bottle" or "Coke" or "diet soda" or "one liter" or "trash."

I occasionally have students read Furnas's paper also; even though it's over 30 years old and refers to out-of-date technology, its conclusions still resonate. I, too, encounter skeptical students. Just like my own classmate 15 years ago, these students find it hard to accept the idea that, even as there could be an infinite number of possible names for most things, there isn't also an especially apt name that is clearly best. After seeing this so often, I am no longer surprised at the persistence of this reaction. But I do find it interesting. Misunderstandings and misapprehensions caused by ambiguous language use are everyday occurrences, even with close friends and family members. But there seems to be a sense among my students that this is a solvable problem for Professionals Thinking Carefully About It. My students want to be those professionals, and it makes them uncomfortable to consider that there isn't some specialized knowledge or magical rite or dataset to analyze that will produce the best name.

The solution proposed by Furnas and his colleagues is to use lots of synonyms. They call this *unlimited aliasing*, but it's just a simple form of controlled vocabulary—quite an old technique, from the perspective of information science. (Controlled vocabularies don't often emphasize unlimited synonyms, but that's because this approach diminishes precision. If all objects are labeled *object*, that word is a spectacularly useless label. Expanding recall at the expense of precision may sometimes be reasonable, but too many results can be a problem when we have billions of everything to select from.)

Controlled vocabularies are a mechanism for standardizing data, rendering it interoperable across data creation contexts. A controlled vocabulary is, in essence, an enumerated set of concepts identified by labels. Often, the labels are words or phrases (e.g., *French bread*), but the labels could be notations or codes (e.g., QA761.2). In its basic form, a controlled vocabulary is simply a list of allowable values. For example, a person filling out an employment survey might be asked to select their current position from a list of job titles. The list of job titles is a kind of controlled vocabulary.

More complicated controlled vocabularies include relationships between the concepts. The simplest relationship is equivalence—Furnas's *aliasing*. A single concept can be associated with many equivalent labels. (For instance, the labels *instructor* and *teacher* might be associated with the same concept.) Another relationship is hierarchical, so that narrower terms are related to broader terms. (A *teacher* might be the narrower term of *educator;* a *teacher* is a kind of *educator*. A *coach* might be another kind of *educator*.) Associative relationships go across hierarchical branches (*teacher* might be related to *school* via an associative relationship; a teacher works at a school, but a teacher is not a kind of school, and so *teacher* is not a narrower term of *school*).

Controlled vocabularies are commonly associated with information retrieval, and they can be implemented in searching even when users are not aware of them. For instance, a controlled vocabulary is in play when you search an online store for *Tylenol* (a brand of pain reliever) and receive *Panadol* (another brand name of the same pain reliever) or *acetaminophen* (the generic name for the pain reliever in the United States) or *paracetamol* (another generic name for the same medication, e.g., in Europe). But controlled vocabularies are used in all sorts of data collection, and the concepts do not have to be represented by words. The Munsell color chart that I referred to earlier, which archaeologists use to describe the hues of soil samples, is a kind of controlled vocabulary in which the concepts are the shades of brown depicted on the chart. Whenever data values are limited to a set of defined choices, a controlled vocabulary is being used.

In a very real sense, such vocabularies don't just affect the data; they *are* the data. Considering the impact they have, it's a little strange that more people don't pay more attention to the details of their construction. Even when problems are identified with vocabularies, it can be difficult to generate a sense of urgency about them. As one example, consider the US census. The census, conducted every ten years, provides baseline population and demographic data that informs both policy decisions (how federal, state, and local governments allocate resources, representation, and so on) and judicial decisions (for instance, when courts determine whether policies align with legal standards for fairness and antidiscrimination). Courts might use census data on race and ethnicity, for instance, to assess whether voting districts have been illegally drawn to dilute the power of Black voters.[10]

Although race and ethnicity data from the census is necessary for this and many other purposes, this data has suffered from increasing lack of precision, as its controlled vocabularies, designed for one social context, are used differently by a changing society. Since 1980, the census has defined "Hispanic, Latino, or Spanish" origin as an *ethnicity* separate from *race*. Respondents are asked one question about Hispanic, Latino, or Spanish origin and a separate question about race (with no option to select Hispanic, Latino, or Spanish as a race). As Margo Anderson explains, in the 1970s, various federal agencies and advisory groups, concerned about undercounts, had pressed the Census Bureau to add a new question about Hispanic origins to the basic census form (Anderson 2015). At the time, none of these groups sought to incorporate Hispanic origin into the existing race question.

Over time, however, racial identifications in the United States have shifted. Increasingly, Latinx people do not identify themselves as White (although they may have done so when these census questions were first added). Now, many Latinx people look at the census vocabulary for race—White, Black, Asian, Native American, Some Other Race—and do not know what to select. Accordingly, in 2000 and 2010, Some Other Race, which was intended to be a tiny category, became the third-largest racial group in the United States (Ashok 2016).

As described on the US Census Bureau Web site, most respondents who selected Some Other Race also identified themselves as Hispanic or Latino, although some respondents of Middle Eastern origin also selected Some Other Race. What does this mean? As with any other data, there will always be some level of uncertainty in the US Census depiction of race and ethnicity. In this case, however, because we have not paid sufficient attention to our controlled vocabularies, census race and ethnicity data has become more uncertain than it might otherwise be.

In 2015, following years of research and testing in response to this situation, the US Census Bureau developed plans to combine the race and ethnicity questions into one and to add new race categories for Latino and for Middle Eastern and North African (MENA). But the Census Bureau must use the race and ethnicity categories approved by the federal Office of Management and Budget (OMB), and the OMB did not grant permission for the Census Bureau to use its new, combined question (with a revised controlled vocabulary for race) in the 2020 census (Wang 2018). This was a political decision; it serves partisan interests if non-White residents are undercounted. However, because this decision had to do with boring, technical matters of controlled vocabularies, it received relatively little media attention. In contrast, an initiative to add a citizenship question to the census was heavily reported in the press. A citizenship question, many feared, would result in undercounts of undocumented populations: specifically, Latinx populations. Ultimately, a series of legal challenges caused the citizenship question to be withdrawn. However, similar consequences may yet result from the decision to retain the old race and ethnicity categories in 2020. The continued prevalence of Some Other Race in the data means that certain populations will be undercounted, particularly Latinx populations.

The Imperative of Data Criticism

The situation with race data in the US Census illustrates an ongoing dilemma. On one hand, reproducibility and interoperability can only be approximated, never achieved, and it's dangerous to imagine otherwise. No data escapes uncertainty completely! So we can't expect too much from reproducible data collection protocols or from technical devices that aim to standardize data by controlling variation. Even with the proposed (but abandoned) race and ethnicity categories for the 2020 census, we can imagine situations in which, as just one example, Americans of Egyptian descent might reasonably identify themselves as White, MENA, or Black.

On the other hand, given that no data can escape uncertainty, we can and do use uncertain data to make good, defensible, so-called *evidence-based* decisions—as long as the uncertainty remains manageable. Accordingly, we still need standardized data collection protocols, controlled vocabularies, and other technical devices, and we need them to work as well as they can. The problem with the old census categories for race and ethnicity isn't that they are merely uncertain; it's that they are highly uncertain. (That millions of people select Some Other Race is but one symptom of this.) The abandoned proposal for 2020 was created to mitigate a known form of uncertainty, making it manageable, within an acceptable tolerance.

But how do we get from highly uncertain to manageably uncertain? If perfect is impossible, how do we know what is good enough? What is the level of standardization, the level of interoperability, the level of reproducibility, that we should strive for? And, conversely, what is the level of standardization, interoperability, reproducibility that we should avoid?

If there were a simple answer to this question, we would be doing it already. This makes our duty severe. We can't just follow rules or rerun some scripts. We have to pay attention to the ebbs and flows of uncertainty in our data and, even as we respect the inevitability of imperfections, we need to make adjustments. In other words: any responsible data *creator*—indeed, any responsible data *user*—also needs to be a data *critic*.[11] We must be critical data creators; we must be critical data users. This critical activity is not an extra add-on to data creation and use, some kind of maintenance step that gets tossed off or forgotten or delegated to the less-equal members of the team. It's part of it.

If anything makes this pill easier to swallow, it is that criticism—good criticism—is not the cruel act that many imagine it to be. No one becomes a critic because they *hate* the object of their criticism; they become critics because they *love* the object of their criticism. That love is what drives a good critic to become an expert. A good music critic has vast knowledge about how music works: about rhythm and melody, about instruments and their tones, about lyrics and singing, about studio production. A good music critic knows about how other people have characterized good music and bad, in different genres, over time. A good music critic learns these things so that they can identify and encourage the growth of new musical forms, to give them loving care.[12]

Similarly, a good data critic knows how data works. A good data critic knows about technical devices like controlled vocabularies—their form and function, as an ideal and as a reality, and how other people have characterized good and bad ones, for different reasons, over time. A good data critic knows about design principles like standardization and interoperability and replicability, and has a reasoned, informed perspective on what those ideas mean, what we can expect from them, and how we can make use of them for data creation and use. A good data critic knows about these things and keeps finding new things to learn about them, and is endlessly curious and captivated, because just like music or movies or video games or cuisine, our understanding of data is just as dynamic and debated.

Importantly, a good data critic isn't trying to hold data to impossible standards of rectitude or to make all data adhere to the same specifications. Like any other form of expression, data has a wholeness that can transcend its individual components, and, accordingly, good data can never just be running through a checklist of so-called *best practices*. If we think about music, it's easy to see what I mean. A great song can have trite lyrics. In fact, sometimes, the banality of the lyrics can *contribute* to the greatness of the song rather than detract from it. Take Marvin Gaye's *Let's Get It On*. As prose, "let's get it on" is not a very enticing invitation. But when Marvin Gaye sings those words, they transform from crude to smooth. We all have our own examples that we can point to here; this is not an unfamiliar phenomenon.

A good data critic knows that a dataset, also, can be more than the sum of its parts. A good data critic doesn't mechanically mark apparent errors, whether they be inconsistencies, inaccuracies, or biases. A good data critic isn't down with conservatism and control. A good data critic looks at the whole—and tries to see what makes it sing.

5
TAXONOMY

This chapter advocates for taxonomic structure as a mechanism for clarifying and communicating semantic relationships between data values. I contend that taxonomies support, rather than inhibit, pluralistic accounts of the world.

In the adventure essay, I tell stories about

- A vile hierarchy.
- The vile hierarchy, revised.
- That jumble drawer in your kitchen.
- Confessions of a data snob.
- "User-centered" taxonomy.

In the reflection essay, I ruminate over

- A constellation of fundamental data concepts.
- Taxonomies and data users.
- Taxonomy development as design, rather than discovery.
- Structural constraint as an invitation to creativity.

ADVENTURE: NOT THAT HIERARCHY

A Vile Hierarchy

Sometimes, when I teach, I try to spice things up with puzzle games. For instance, I might display an organizing system that I've recently encountered, and then ask the students to explain it to me.

Here's a fun one. It's an excerpt from a taxonomic structure used to organize an art database.[1]

```
Gestures
    Hand
        Head resting on hand
        Hiding or finding a treasure
        Hitting
```

```
Horror & Fear
Killing
Levitating
Lying on a cloud
Nymphs
Obscene
```

The gestures in this list are *potential values to create data* about the artworks. For instance, if we had a painting of the annunciation, where an angel is holding a hand out toward Mary, then we might describe that painting as having a Hand gesture (see figure 5.1).

This list of gestures is vile! What an awful hierarchy! It makes my blood boil.

What's wrong with it? This is the puzzle.

It's not supposed to be a trick question. But when I first started playing these games in class, the students would often respond like this:

> These categories are biased and exclusionary. They suggest a society that promotes violence, exalts wealth, objectifies sex, and takes an individualistic perspective on everything, even religion. It's a caricature of Western patriarchy in capsule form.

To which I would reply, "Splendid thematic analysis! Great work! But what about the hierarchy?"

And then there would be silence.

Argh! What had happened?

Maybe you are also puzzled. It took many such encounters for me to understand what was going on.

Gradually, I realized that when I, as a classificationist, talked about *hierarchy*, I meant *taxonomic relations*. But when the students talked about *hierarchy*, they meant *power relations*. Not that hierarchy at all!

What's the difference? In hierarchical power relations, the classes at the top of the hierarchy have qualities not shared by the classes beneath. For instance, in a monarchy, there is an inherent and insurmountable difference in power between a queen and her subjects. We might diagram such a hierarchy this way:

```
Queen
   Nobles
      Serfs
```

The queen rules over nobles, who rule over serfs, who rule over no one.[2] In a power hierarchy like this, the qualities of interest—noble ancestry and the privileges associated with it—get diluted on the way down. The queen

Figure 5.1
To describe this painting, we might assign the Hand value from the Gestures taxonomy. (Alesso di Benozzo, The Annunciation, ca.1480–1500. Metropolitan Museum of Art, New York. Robert Lehman Collection, 1975.)

has the most direct ancestry and the most privileges. The nobles have some connections to the royal line, and some privileges. But serfs do not share in either of these qualities.

When my students interpreted "vile hierarchy" as a reference to power, they were identifying the unsaid power relations that condition what we consider to be art, what kinds of gestures appear in art, and how we describe those gestures. Hierarchical power relations, in other words, inform the thematic coherence of the gesture categories.[3]

In contrast, hierarchical relations in a taxonomy involve the inheritance of qualities from the top of a tree downward. Whatever is true of the top of a taxonomy is also true of everything beneath it, and new qualities are added, rather than subtracted, at every level. In a taxonomy, the classes at the *top* are comparatively impoverished, while those at the bottom are comprehensively specialized.

For instance, in a taxonomy of leavening agents, every class shares the quality of trapping air in a dough or batter; this is a fundamental similarity between all leavening agents. Here's a simple hierarchy that shows this kind of inheritance:

```
Leavening agents
    Air
        By creaming (solid and sugar)
        By whipping (liquid and sugar)
            Egg whites (animal product)
            Aquafaba (vegetable product)
    Steam
        Via lamination (steam produced by layered butter in
        dough)
        Via steam as cooking method
    Carbon dioxide
        Chemical
            Sodium bicarbonate (baking soda)
            Sodium bicarbonate plus acid (baking powder)
        Biological
            Wild yeast
            Baker's yeast
```

According to this taxonomy, yeasts are Leavening Agents that trap air by means of releasing Carbon Dioxide into the dough or batter via Biological Processes. We could alternatively say that baker's yeast is *a kind of* leavening agent, and anything that is generally true of leavening agents (i.e., that they trap air) is true of baker's yeast. Aquafaba and baking soda are equally kinds of leavening agents.[4] This is the opposite of queens and serfs: serfs are *not* a kind of queen.

Another distinguishing characteristic of taxonomic hierarchies is that all the classes at a given level are related to their parent by *a single principle of division*. For example, at the first level in the taxonomy of leavening agents, the principle involves the means by which air bubbles are introduced into the dough or batter. Air is introduced directly (Air), or through Steam, or through Carbon Dioxide. When air is introduced directly, it is done by one of two methods: by creaming (which involves beating a solid fat and sugar) or by whipping (which involves beating a liquid and sugar). The principle that relates Air to its subclasses, then, involves what is being beaten (a solid or a liquid).[5]

My students, thinking of hierarchy in terms of power relations, focused on the *thematic coherence* of the categories I showed them. But I was thinking about hierarchy in terms of taxonomic relations, and I was focused on *structural coherence*. Structurally, the set of gesture categories is vile because it involves many different principles of division at a single level. Some of the categories involve the parts of the body used to make the gesture; some of the categories involve a direct physical action (like lying down); other categories involve a functional action that can be conducted via various physical means (like killing); other categories involve reactions (like horror) . . . well, it's a mess, from a taxonomic perspective. One could argue that it's not a hierarchy at all.

Although hierarchy in the sense of taxonomy is structurally quite different from hierarchy in the sense of power relations, it is common to view them as conceptually similar. In particular, any form of hierarchy tends to be seen as operating within the same system of ethical values, a system that perpetuates inequality. If our politics support more democratic and inclusive power arrangements, we often instinctively oppose hierarchy in all forms. According to this form of reasoning, a flat structure is better than a hierarchical one.

However, as my students' analysis shows, the example list of gesture categories exhibits thematic alignment with a power hierarchy, even as it violates basic principles of taxonomic structure. If that can happen, then the two forms of hierarchy must be at least partially disjoint. In fact, I would argue, data that lacks structural coherence—data that is not hierarchical in a taxonomic sense—can indicate that our data is riddled with exactly the kinds of assumptions that we might, as advocates for democratic values, want to surface. In short, data that does not align with taxonomic principles is hiding something.

Both power hierarchies and taxonomic hierarchies do, certainly, instantiate a specific, limited perspective on the world. However, they express this perspective differently. Hierarchical power relations rely on an unequal distribution of power from the top of the hierarchy to the bottom. A single perspective on the world—a perspective that validates this allocation of power and the corresponding social roles required to support this arrangement—is necessary to maintain the integrity of the system. What the queen says, goes; the serfs are presumed to accede to the queen's will.

(Should the serfs think differently, their perspective is disregarded as contrary to the proper order.)

Taxonomies, too, enact a single, limited perspective. Taxonomies are driven by a particular characteristic: for instance, leavening agents by their mechanism for introducing air bubbles (and not by their smell), plants by their morphology (and not by their agricultural uses), or living organisms by their evolutionary descent (and not by their commercial value). Furthermore, each level of a taxonomy can involve only one organizational principle at a time. Leavening agents that introduce carbon dioxide are subdivided only by whether they are living organisms (biological) or inorganic substances (chemical). As with power hierarchies, a single perspective is therefore enforced, and others are silenced. My taxonomy of leavening agents focuses on how the air bubbles in dough are created. It doesn't say anything about the taste or texture that might be associated with a particular agent. From the perspective of my taxonomy, those characteristics don't matter, even as those characteristics might be quite important for bakers.

Both power hierarchies and taxonomic hierarchies, then, view the world through a partial, directed lens. However, where a power hierarchy *obscures* its rationale, a taxonomy *clarifies* its rationale. A power hierarchy typically requires exclusivity; it cannot coexist with other systems at the same time. Accordingly, power hierarchies gain legitimacy by appearing to exist naturally, without alternative. A power hierarchy that is documented can be debated, and, if you are the queen, that's where the trouble starts. It is to the queen's advantage to assert her position, rather than explain it.

In contrast, where a power hierarchy provides no argument to validate its structure (it just assumes its premises), a taxonomy's argumentation is, *by design*, open to inspection. If a taxonomy is well constructed, it is structurally transparent. Because each level operates by means of a single principle of division, we can systematically trace relationships between classes. If the structural logic is poor, deficiencies can be identified through rational argument. For instance, my taxonomy of leavening agents asserts that air is incorporated into doughs via three mechanisms and no others: via air directly, via steam, and via carbon dioxide. If my taxonomy is well made, then the three mechanisms of incorporating air are *jointly exhaustive;* in other words, no other mechanisms exist. This is a direct claim that the taxonomy is making. If this claim is mistaken—if there is another type of mechanism for introducing air that I have missed—then my taxonomy is wrong and you can tell me precisely why that is (it's wrong because the subclasses aren't jointly exhaustive).

Similarly, if my taxonomy is properly made, then the three mechanisms of incorporating air are *mutually exclusive;* in other words, there is no leavening agent that introduces air by multiple mechanisms. Again, this is a direct claim that the taxonomy is making. Other claims logically follow. For example, because egg whites are in the Air branch, my taxonomy claims that whipped egg whites *do not* introduce carbon dioxide into a dough. If this claim is mistaken—if some leavening agents work via multiple mechanisms for incorporating air—then my taxonomy is wrong and you can tell me precisely why that is (it's wrong because the subclasses aren't mutually exclusive). I would then need to adapt my taxonomy in light of this new information. If I can't (or won't) adapt it, well, then the flaws in my argument remain out there in the open, for all to see and judge. In this way, a carefully constructed taxonomy invites one to consider its flaws and develop alternatives.

The Vile Hierarchy, Revised

However, like the set of gesture categories for describing artworks, most of the taxonomies that we encounter in everyday life are not well constructed. These taxonomies hide their workings and motivations just as power hierarchies do. Such taxonomies employ multiple principles of division when relating a parent class (like Gestures) to its subclasses (like Head and Hitting). As a result, subclasses are not, and can never be, jointly exhaustive and mutually exclusive. The reason is that they mix many different kinds of characteristics at once (such as the body part used to perform a gesture [Head] and the purpose of the action represented by the gesture [Hitting]). These characteristics might be thematically similar (they all have to do with gestures and, as my students would likely point out, they are all informed by the same set of social values, which align with a certain power hierarchy), but they are structurally dissimilar.

When data lacks structural coherence, it can be challenging to aggregate and compare. An analogy might be taking a bunch of measurements of a person's body, such as overall height, inseam, torso length, circumference of the head, circumference of the waist, and so on, and then mixing them all together without saying what they refer to. You've got a lot of data—for example, 173 cm, 84 cm, and 51 cm—but all you know is that they represent some set of body measurements. Moreover, each person that's been measured might have a completely different assortment of values (one person has the inseam and circumference of the wrist, another person has overall height, hip circumference, and neck circumference). These measurements are thematically similar but structurally dissimilar, and it's hard to make sense of them. In contrast, if we were specific about each kind of measurement taken, instead of dumping all the measurements together with no differentiation, we would be better able to make sense of the measurements across different people.[6]

In trying to understand what a dataset leaves unexpressed, it can be insightful to take a set of data values that appear thematically coherent and see how we might relate them structurally. The more hidden gaps that lurk in the initial value space, the more work that we have to perform to implement structural coherence. For example, if we take a list of potential data values such as the gesture categories and attempt to apply taxonomic structure to it, we have to add a tremendous amount of missing logic, in the form of layers and layers of subclasses in multiple parallel hierarchies. The missing logic required to relate the data values represents a silent infrastructure of hidden assumptions.

Thinking taxonomically, I might restructure the gesture categories this way (the bold terms are the values from the original list):

```
Gestures
    <Gestures by the body part used to perform the gesture>
        <Gestures that use the whole body>
        <Gestures that use parts of body>
            (Gestures that use the) Head
            (Gestures that use the) Hand
    <Gestures by purpose of the action represented by the
    gesture>
        <Gestures that represent purposive actions>
            <Gestures that represent actions toward people>
                <Gestures that represent violent actions>
                    (Gestures that represent) Hitting
                    (Gestures that represent) Killing
            <Gestures that represent actions toward objects>
                <Gestures that represent locating an object>
                    (Gestures that represent) Hiding
                    (Gestures that represent) Finding
        <Gestures that represent reactions>
            <Gestures that represent emotional reactions>
                (Gestures that represent) Horror and fear
```

```
<Gestures by their alignment with social codes of
conduct>
   <Gestures that are socially acceptable>
   <Gestures that are not socially acceptable>
      Obscene gestures
<Gestures by the objects being gestured toward>
   <Gestures toward living objects>
   <Gestures toward inanimate objects>
      <Gestures toward valuable objects>
         (Gestures that are directed toward) Treasure
<Gestures by the type of being performing the gesture>
   <Gestures performed by humans>
   <Gestures performed by supernatural beings>
      (Gestures performed by) Nymphs
<Gestures by location of the gesturer>
   <Gestures performed on land>
   <Gestures performed on water>
   <Gestures performed in the air>
      (Gestures performed on) Clouds
```

Kind of overwhelming, isn't it? Bear with me, and I'll explain what I've done.

Everything in brackets (< >) represents a conceptual gap that I had to fill in order to clarify the structural relationships between the different data values. Seems like a whole lot of brackets, right? I actually kept it sparse![7] But the scariness of all those brackets demonstrates my point. Those brackets represent unstated assumptions. The thematic coherence of the data values was hiding an immense amount of missing context. Primarily, this missing context involves the kinds of data being collected; in other words, different aspects of the gesture are being described in the data (the physicality of the movement performed in the gesture is, for instance, a different kind of concept from the object that might be gestured toward).

Still overwhelmed? What, you might well ask, is this *Gestures by* business? The first level of the new taxonomy is all *Gestures by* something or other. Good eye! Strictly speaking, this is a kind of technical cheat. The original gestures list brings together many different aspects of a gesture; in other words, it uses many different principles of division. Creating a single hierarchy from this list would be unmanageably complicated. And so we cheat a little. We specify multiple parallel hierarchies, one for each aspect of gestures that we are interested in. This is called a *faceted* classificatory structure.[8] Each facet represents a different kind of data, or a different aspect of the gesture (like the facets on a jewel). Within each facet, we can proceed on the basis of normal taxonomic principles.[9]

If you've been puzzling over all those brackets, perhaps you disagree with the decisions that I've made, thinking that you would arrange the taxonomy differently. Yes! Certainly you could. One can think of taxonomic principles as similar to rhyme and meter conventions in poetry. There are lots of ways that you can create a taxonomy that includes certain data values, just as there are lots of ways that you can write a sonnet that includes certain words. This is what I mean when I suggest that taxonomic structure is not oppositional to plurality and inclusiveness. A well-constructed taxonomy operates according to a systematic form of discourse; this is what makes it an *argument*. When you design a taxonomy, you take a position (by selecting characteristics with which to describe things) and you present evidence to substantiate that position (by systematically subdividing those characteristics according to a single principle of division per level). Someone else can always take a different position and present different evidence. The point of developing a taxonomic structure isn't to be right; it's to clarify one's assumptions and demonstrate the flow of one's argument, in a manner that can be easily read.

A common reaction to *all those brackets* would be that I took a simple list and made it unbelievably more complicated. But I didn't! That "simple list" of original gesture categories was *deceptive*. It was hiding its complexity. My taxonomic analysis was a form of critical intervention: I revealed some of the unstated assumptions that lurked within the simple list.

Note that those brackets in themselves don't *do* anything to the actual data created with the gestures list. That is, if an Edo period woodblock print showing an altercation between two disheveled gentlemen was previously assigned data values of Hand and Hitting, a taxonomic analysis doesn't change that data (see figure 5.2).

The taxonomic analysis does, however, help us to interpret the data, to better understand what it represents. The taxonomic analysis (as indicated by the brackets) identifies where the flat structure of the gestures list was hiding implicit structural relationships in the data values. The brackets clarify that Hand and Hitting are different *kinds* of data values that represent two perspectives (or facets) on the gesture in the painting. In the print, the beardless man on the right *is making the gesture with* his hand, reaching back to strike the man on the left; *the action represented by* the hand is *hitting*.

Further, this depiction of structural principles can help us to think more carefully about how to implement the existing data design. If we consider what the taxonomic analysis tells us about the different kinds of data values

Taxonomy 139

Figure 5.2
To describe this woodblock print with the Gestures taxonomy, we might assign values of Hand and Hitting. (Utagawa Kunisada, untitled print, 19th century. Metropolitan Museum of Art, New York. Gift of the Estate of Samuel Isham, 1914.)

rolled up in the current gestures list, then what we should we actually be collecting for each artwork that we describe, if we want our data to be coherent and consistent across artworks? We have seven facets, which means that the data design comprises seven kinds of data for each gesture. So, does that mean that when we're describing the gestures in a work of art, we should make sure to collect each of the seven kinds of data, systematically and comprehensively, as they apply to each work? Or, on the other hand, do we need to reevaluate our data design? Do we need seven kinds of data about gestures? Are these seven kinds the right ones? Maybe, instead of describing seven perspectives on gestures with a single flat list of potential values, we could describe gestures with three distinct attributes, the Purpose of the Action Represented by the Gesture, the Part of the Body Performing the Gesture, and the Being Making the Gesture. (Or five attributes. Or three. Or twelve. But you get my drift: understanding the structure of the existing design can help us to envision new possibilities.)

That Jumble Drawer in Your Kitchen

Okay, be honest. Have you fallen asleep in a puddle of drool?

You wouldn't be the first. As much as I find them fascinating, the actual mechanics of data infrastructure are stupendously boring to most people. But even if I lost you at the word *facet*, my repeated invocation of these diagrams and details should at least demonstrate my commitment to the power of taxonomic principles. In fact, you could call me a taxonomic evangelist.

I do believe! I believe that structural coherence can help us understand data more perceptively, interrogate it more critically, and design it more robustly. Thinking taxonomically encourages us to focus on *relationships between* data values, rather than individual values. Moreover, thinking taxonomically demands that we systematically articulate the nature of those relationships in a way that humans can understand (as opposed to, say, the machinations of neural nets, deep learning, and other forms of algorithmic patterning whose arguments aren't meant to be comprehensible to human beings). A hand makes a gesture, but the act of hitting is depicted in a gesture. When we think taxonomically, we can't hide in the misleadingly flat environment of the list, where we can throw all kinds of different concepts together (Hand, Hitting, Nymphs, Obscene?!) and not even realize what we've just created: data infrastructure that's exactly like the drawer in your kitchen where you have scissors, playing cards, shoelaces, a hammer, city maps, matches, and spare packets of soy sauce and ketchup from meals of takeout past, among other things. Oh my, that jumble drawer in the kitchen! I bet you don't even know what's in that drawer.

So, yes. I am a taxonomic evangelist. Hear me preach! Responsible data design doesn't *avoid* hierarchy, my friends. Responsible data design *uses* hierarchy. Taxonomic structure is a form of explanation; in laying out assumptions about how data values relate, taxonomic structure facilitates critique.[10] Cue the chorus of angels!

So. A humble servant of the taxonomic gospel. That's me.

Or . . . there is another, less noble interpretation of my predilections.

I am a data snob.

When you are a *snob* of something, you believe that you can separate good specimens from poor ones. What's more, as the snob of something, you believe that you know what *good* is.

For instance, I am a tea snob, in addition to a data snob. As a tea snob, I believe that tea is *always* better when boiling water is poured onto loose tea leaves in a teapot. Tea made by dunking a bag into a mug of hottish water is *not as good* as properly brewed tea. As a tea snob, I likewise assert that drinking temperature is incidental, even as brewing temperature is vital. Personally, my own tea must be scalding when I drink it. But a tepid cup would be of equal quality. The point is that as a tea snob, I know which characteristics matter and which don't.

If we reinterpret my enthusiasm for taxonomic principles as a form of data snobbery, my position doesn't change, but perception of it might. We might respect an evangelist's purity of faith, even if we are not convinced. But who respects a snob? Snobs are petty, self-serving, and status seeking. If I'm a data snob, then my subconscious motivation is not better data but data that I am better at. (There is no need for modesty; I am a *sick* taxonomist. Of course I want to align taxonomic structure with data quality.)

Confessions of a Data Snob

Maybe I am just a data snob. After all, principles of division remain obscure in most of the everyday data that we encounter, and few people seem to notice. Take online shoe shopping, which often employs a faceted browsing structure. Theoretically, each facet represents a single characteristic to differentiate shoes (e.g., color, price, or size). Often, however, the potential data values within a facet combine multiple principles of division. For instance, on shoe sites, it's common to see a Color facet like this:

```
Color
    Black
    Dull
    Metallic
    Navy
    Ombre
```

All these values have something to do with color, but in different ways. Some values describe hue (Black), while others describe shine (Metallic), saturation (Dull), or pattern (Ombre). From a taxonomic perspective, we'd clarify what we meant by Color if we added another level of hierarchy like this:

```
Color
    Hue
        Black
        Navy
    Pattern
        Ombre
    Saturation
        Dull
    Shine
        Metallic
```

The primary facet in shoe data presents a particularly vivid example of unclear taxonomic principles. For different online retailers, this facet might be identified as Category (as in Shoes by Category) or sometimes Style or even just Shoes. But what kind of data goes here? It's hard to explain.

Table 5.1 displays data values for this facet from a few popular online retailers for women's shoes.

These flat lists combine many principles of division. Most involve permutations of form: how the sole is attached to the shoe (sandals are attached by straps, other shoes are attached by larger pieces of material); whether the shaft of the shoe extends above the ankle (boots); how the shoe is fastened (laced up like Oxfords, or slipped on like loafers). Other kinds of data values in this facet involve the following:

- Heel height (e.g., high heels, which can align with any form).
- Heel shape (e.g., wedges, which can also align with any form).
- Where the shoes are worn (e.g., slippers, which are worn inside only).

Taxonomy

- The specialized function or purpose of the shoes (e.g., athletics, which are worn to play sports).
- The social situations where the shoes are appropriate (e.g., dress shoes, which are worn in formal situations).
- The style of the shoe or how it looks and fits (e.g., fashion sneakers and boat shoes. These two styles were once associated with specialized functions but have transcended those functions. Today, most people wearing Chuck Taylors aren't playing basketball, and most people wearing Top Siders aren't steering a yacht).

This shoe data obscures a number of relationships between the values. For instance, Oxfords and loafers are both closed shoes (not sandals) that do

Table 5.1
Data values from the primary facet for three online retailers of women's shoes, as of July 2020. The data values are similar across the three sites, but each retailer labels this facet differently: Category, Style, or just Shoes.

Zappos Women's Shoes by *Category*	Onlineshoes.com Women's Shoes by *Style*	Amazon Women's Shoes by *Shoes*
• Boots • Sneakers and athletic shoes • Sandals • Heels • Flats • Loafers • Clogs and mules • Slippers • Oxfords • Insoles and accessories • Boat shoes	• Boat Shoes • Boots • Dress • Flats • Heels • Lace-up • Loafers • Mary Janes • Oxfords • Running • Sandals • Shoes • Slip-ons • Sneakers • Wedges • Work • Barefoot • Pull-ons • Slip-ons	• Athletic • Boots • Fashion sneakers • Flats • Loafers and slip-ons • Mules and clogs • Outdoor • Oxfords • Pumps • Sandals • Slippers • Work and safety

not extend over the ankle (not boots). They differ in fastening: loafers do not have fasteners, but Oxfords have laces. Sandals, in contrast, differ from all the other shoe forms, in that only sandals have the sole attached by straps. In other words, Oxfords and loafers are more specific categories than sandals and involve more kinds of formal characteristics. Accordingly, Oxfords and loafers are more similar to each other and less similar to sandals.

From the perspective of my data snobbery, all the shoe facets in table 5.1 are structurally incoherent. If you are not a data snob, on the other hand, you are probably still struggling to comprehend exactly what's so appalling about putting sandals, loafers, boots, and sneakers together in a single list. If this is so dreadful, then why does it seem normal?

In everyday life, we regularly use concepts like boots and sneakers without being able to define and relate them taxonomically. Here's a question: are high-top sneakers *boots*? I've asked this question during lectures, and most people say no, high-top sneakers are not boots. It would be wrong, these people say, to browse through boots on a shoe-shopping Web site and find a bunch of high-tops mixed in. At the same time, it is difficult to explain exactly why this should be the case. High-top sneakers and boots share a basic form. Most boots are made of leather—but so are some high tops. Most high tops are fastened with laces—and so are some boots. In another era, rubber soles were found only on high tops. But rubber, PVC, and similar materials are used in all sorts of shoes today, across use contexts and levels of formality. In sum, a black leather high-top sneaker and a black leather lace-up boot are quite similar in formal aspects. In a taxonomy based on formal characteristics, they would be in the same place. The difference between boots and high-tops is either stylistic or functional: high-tops are boots with athletic styling (e.g., "fashion" sneakers) or an athletic function (e.g., shoes actually worn to play basketball). Taxonomically speaking, the relationship between boots and high-tops is associative, meaning that they operate on different principles of division, or different facets.

Our everyday concepts are riddled with these kinds of taxonomic gaps. Loafers and boots are both narrower, more specific concepts than sandals. But we're not generally aware of this, because we don't have a ready-to-hand label to identify an intermediate class that includes both loafers and boots (i.e., "all the shoes that do not attach the sole by means of straps"). In fact, if we made up a name for this concept, such as "closed shoes," or "shoes that are not sandals" and put that in the shoe facet, that would be confusing to most people, even if it might close some taxonomic holes. Surely, only a trifling snob could endorse a data-design approach that prefers an artificial structural coherence to the concepts that people actually use. Right?

"User-Centered" Taxonomy

Indeed, when I assign taxonomy design projects, my students often adopt this stance. When I first teach students about taxonomies, and I show them examples like the shoe facets, they are aghast. That data is appalling, they agree. But when they try to design their own taxonomies, they get uncomfortable. They want to believe that data design, like all design, should be "user-centered." They imagine that data design is just documentation: that data designers discover and record our common understanding of sandals, loafers, boots, and sneakers. But if data design is just a strict translation of "what users think," then a data designer shouldn't "make up" concepts that don't actually exist, like "closed shoes," merely to illustrate a taxonomic progression that people don't actually notice. If the data designer is making things up, which taxonomic thinking seems to require, then that gives the designer too much power. My students, typically, do not want that power.

My students feel even more uncomfortable when the design process reiterates that taxonomic structure, because it is a single limited perspective, entails exclusion. A data design that follows taxonomic principles will not include boots and sneakers in the same facet, because boots are distinguished by formal characteristics, whereas sneakers are distinguished by stylistic or functional characteristics. To design a taxonomically sound data structure that incorporates both boots and sneakers, we need multiple facets. But a data design that separates boots and sneakers seems weird and awkward, because this isn't a distinction that people *naturally* make. If you were to ask people to list types of shoes, those lists would likely include both sneakers and boots. Indeed, such a list would probably look a lot like the original, nontaxonomic shoe facets that I said were atrocious. When my students assimilate that atrocious data is also what seems natural, they recoil from taxonomic thinking. It seems snobbish, now, to assert that good data needs structural coherence, when everyday usage is taxonomically chaotic. My students don't want to be snobs.

I understand the students' reluctance. Data structures have power, and when a data designer enacts structural coherence, the designer's role in focusing that power becomes apparent. This is scary. It's scary to admit that you've created a design based on *your* idea of what good data should be.

So, I get it. But data designers *always* have that scary power. It's merely that, with taxonomic structures, the designer's role is more visible. I know that I just said that if you were to ask people to list types of shoes, those lists would likely include both sneakers and boots. And that's probably true. But there would still be an incredible amount of variation between such lists, because we don't actually have a common, orderly way of thinking about such things. The idea of "concepts that people actually use" is a mirage. Even the most unadorned documentation of *what people supposedly think* involves an absurd number of design choices. That's why those online retailers have different shoe facets, after all. In other words, although a data design intended to replicate what people actually think might seem "more natural," it's just as artificial as any other design and just as much a potential tool of exclusion. Remember what I said earlier? If any data is hiding a power hierarchy, it's usually data that seems the most natural. Any sense of greater naturalness, or greater inclusivity, is a sham. You could even call it dishonest.

A data structure that aligns with taxonomic principles, in contrast, is forced to make some attempt to explain itself. One can see the imposition of hierarchical structure as the data designer's interpretive rationale laid bare. Any particular taxonomy might espouse views that I agree with or views that I abhor; it might make sense to me, or I might find its argumentation specious. But it gives me more to argue with; it's more honest. Yes, it might seem initially weird and awkward to introduce "made up" concepts like "closed shoes" in order to demonstrate the progression of characteristics that relates sandals to boots, Oxfords, and loafers. But this move attempts to explain, rather than obscure, what the data means. I would call this data more honest.

When it comes to online shoe shopping, one might still argue that the appearance of honesty is sufficient—in fact, better. After all, it's the natural appearance of those unwieldy shoe facets that keeps customers with short attention spans from abandoning an online boutique. Imagine that we separated boots and sneakers into different facets: "I can't find the sneakers," would say the impatient shoppers. We've likely all been that impatient shopper at one time or another, the person who just wants to see the cross-training shoes already, grrr. But let's be honest here: this design is, in its own mundane way, promulgating a false truth for profit.

"Oh, lighten up already! Let us buy our sandals and boots in peace, without undue thinking about what goes where! Stop it with the lectures, snob!"

I'll stop the lecture. And I *am* a data snob. But mine is the way of the righteous, and maybe you should be one, too.

REFLECTION: HOW DOES CLASSIFICATORY STRUCTURE MATTER?

A Constellation of Fundamental Data Concepts

What is a taxonomy and what does it have to do with data?

To answer this question, I'll need to back up. You see, it would be wonderful if we had a common vocabulary to talk about the structural aspects of data: a set of terms that everyone understood and agreed with.

But we don't.

Certain words pop up with frequency—*ontology*, perhaps, or *schema*, or *metadata*—but none of these has a consensus definition. Worse, some people assert a very specific meaning for a certain term, and if you do not use this term in the prescribed way, you are wrong. Meanwhile, others will use the same words in a looser, expansive way, and complain that the first people are misguided blowhards. Basically, whenever you see a word like *ontology* or *metadata* or even *data* itself, you can't be sure what anyone is talking about.

With that in mind, here's how I understand the fundamental concepts at play here.

Data is just a description. Sometimes data is *unstructured*, or freeform. I could write a paragraph to describe the apple on my desk, or I could list a set of keywords about it (yellow, crispy, 200 grams), or I could use Google to search the Web for *Topaz apples*, and Google could record my query terms and the subsequent clicks that I made. All these descriptions could be considered a kind of unstructured data.

Sometimes data is *structured*; it has a defined form. One kind of structure is logical, or classificatory. This involves separating our description into conceptual parts, so that we systematically collect data for specified characteristics, such as the Color, Texture, and Variety of an apple. Common synonyms for *characteristic* include *attribute, element*, and *property*. A set of characteristics to describe a particular thing (like Color, Texture, and Variety for apples), might be called a *schema*. The logical structure of a schema might be *encoded* in different ways. For instance, a schema might be encoded as a set of tables in a relational database, or it might be encoded as a set of statements in a graph.[11]

The *elements* (characteristics, attributes, properties) in a *schema* define what the data might be. The data itself happens when we describe some thing according to the elements: when we say that the Color of a particular apple (like the one on my desk) is Yellow. *Yellow* is a data *value* for the Color *element*. With structured data, we often constrain the possible data values in some way. There are many kinds of constraints (sometimes called *parameters*). We might say that the data must be in a particular format (such as text) or that it must follow a particular syntax (such as an integer followed by a unit of measurement, for a Weight). We might also specify a particular set of possible values and no others: the Color of an apple can be Red, Green, or Yellow, with no other possible values. This set of possible values is called a *controlled vocabulary*.

Controlled vocabularies can be very simple, or they can be very complicated. For instance, a controlled vocabulary of apple varieties could be quite large. In order to manage the large number of apple varieties, we might group them into categories, such as Eating Apples and Cooking Apples. A Topaz apple is an Eating Apple; a Northern Spy apple is a Cooking Apple.

```
Apple varieties
   Eating apples
      Topaz
   Cooking apples
      Northern Spy
```

When we organize the controlled vocabulary like this, proceeding from more general concepts to more specific ones, we are giving the controlled vocabulary a hierarchical structure. This structure is also called a *taxonomy*. (We could equally just call this structure a *classification*, or even an *ontology*, if we wanted to sound fancy.)[12]

So in the world of data, a taxonomy is a structured kind of controlled vocabulary—a tool that defines *the potential values that data might have*. We use these controlled vocabularies to create data.[13]

Taxonomies and Data Users

Sometimes, we also use taxonomies and other controlled vocabularies for *data access:* to arrange or organize a dataset so that we can better browse it, retrieve items from it, or understand its character. The challenging bit here is that a data user doesn't usually interact with a taxonomy directly, but

Taxonomy 149

rather with a particular dataset that has been organized or arranged according to the taxonomic structure. Let's take a simple example: lunch at the Law Faculty canteen at the University of Copenhagen. The food available for lunch at the canteen aligns with a very simple taxonomy:

```
Lunch items by temperature
   Hot food (the "dish of the day")
   Cold food
      Sandwiches (cold food on top of bread)
      Salads (cold food not on top of bread)
```

The canteen serves lunch buffet style. You get a plate and then take however much of as many dishes as you like. There is a certain consistency in the available offerings, but the actual items vary each day. (On Monday one might find kale salad with apples, and on Tuesday radicchio with blue cheese. Or one day there might be corn to put on your salad, but no peas; on the next day, peas but no corn.) The dishes are arranged on three tables: one table for hot food (let's call it table 1, to the left in figure 5.3) and two

Figure 5.3
The hierarchical organization of the lunch canteen in the Law Faculty at the University of Copenhagen. Table 1, to the left, has hot food; table 2, at the back right, has cold food (sandwiches); and table 3, at the front right, has cold food (salads).

tables for cold food. One of the cold food tables is for sandwiches (table 2, at the back of figure 5.3). The other cold food table is for salads (table 3, at the foreground in figure 5.3).

The employees who work at the canteen take each lunch item and place it on one of the tables. The employees, here, are acting as data creators. When a canteen worker takes a bowl of roasted vegetables and puts it on table 3, this is like assigning the bowl of vegetables the data value of Cold Food: Salads. When a canteen worker takes another dish of roasted vegetables and puts it on table 1, this is like assigning the second dish of vegetables a data value of Hot Food. And when a canteen worker takes a tray of roasted vegetables on pieces of rye bread and puts it on table 2, this is like assigning the tray of vegetables a data value of Cold Food: Sandwiches.

When you go to get your lunch, you don't interact with the taxonomy itself. The tables don't have signs on them that say Sandwiches or Salads or Hot Food. You simply interact with the lunch items, which have been grouped together based on the data values assigned by the canteen employees: sandwiches together, salads together, and hot food together. As a lunch goer, you might remain ignorant of the taxonomic structure that underlies your experience at the canteen, in that you might never consciously recognize that the items have been divided this way, or that the Sandwiches and Salads are two subclasses of Cold Food. But after you go to the canteen once or twice, and you want salads for lunch, you might find yourself going directly to table 3 and skipping the other tables entirely. When you do that, you'll feel pretty confident that all the salad options will be available for you on table 3. There won't be some random beets hiding on table 1, or a bowl of kale salad on table 2; neither will there be a steaming dish of stew on table 3. In other words, even if you don't know that there is a taxonomy underlying the arrangement of lunch dishes, you might be taking advantage of taxonomic principles to facilitate your interactions with the dataset (here, the "dataset" is the placement of each dish on a table).

Taxonomy Development as Design Rather Than Discovery

Wasn't that exciting, that detailed reading of lunch tables? When I've asked colleagues to read drafts of this chapter, they tend to groan about the examples. Gestures, leavening agents, shoes, lunch tables? Gah. But I don't think it's the content. There's just something about taxonomical structure that makes any subject matter seem boring. When the sociologists Geoffrey Bowker and Susan Leigh Star seek to elevate the study of "boring things," they invoke a classification scheme (specifically, the International Classification of Diseases, or the ICD) as the epitome of dullness. The ICD is used to create data about causes of death, which should be very exciting![14] But in classification form? Zzzzz.

It might be surprising then, to consider that developing a taxonomy is intensely creative. The classificationist, like any writer, needs to play with the concepts in the taxonomy, figuring out how to break them down in the most insightful, elegant, balanced way. Moreover, the design process inevitably reveals previously unknown features of the domain, along with the correction of previously unrecognized fallacies and assumptions. For instance, when I created my tiny taxonomy of leavening agents for the adventure essay, it took some ingenuity. I hadn't previously considered the difference between creaming a solid and whipping a liquid, and I hadn't been aware of the leavening properties of the liquid in a can of chickpeas (aquafaba). Nor was the conclusion obvious from the start. It was an artistic experience more than anything else, like Cézanne stepping back from the fruit bowl and seeing in terms of basic shapes: the round solidity of the apples.

For most readers, though, the taxonomy of leavening agents seems to take everything that might be interesting out of baking bread and cakes and drive it far away. I suppose it's because a taxonomy's expression appears so systematic and regular, and its level-by-level construction seems so very painstaking. I find this instinctive reaction unfortunate. What might appear as bland matter-of-factness is really an expression of artistic sensibility. We're just not used to conceiving of it that way.

It's true that, historically, taxonomy has more often been viewed as a process of scientific discovery rather than one of artistic creation: the grouping of things according to their essences, to establish an order that is stable and universal rather than dynamic and contingent. The pull of this idea remains strong to many. In scientific practice, however, it has been convincingly repudiated.[15] Biological taxonomists, for instance, agree that living organisms have no magic essence that distinguishes one species from another (Mayr 1988; Dupré 1993).[16] Furthermore, there are multiple, competing schools of thought regarding how species should be identified, grouped, and related into larger taxa. Some philosophers of science have, moreover, argued that such pluralism in taxonomic approaches strengthens science (Bryant 2000;

Ereshefsky 2007). This doesn't mean that scientific taxonomy lacks evidence, only that such evidence can be interpreted in multiple ways—just like we can group leavening agents in multiple ways.[17]

The notion of a true order of knowledge has also had a powerful influence on conceptions of taxonomy (or classification, as it is more often called) within information science and its precursor, librarianship.[18] Unlike biological taxonomists, bibliographic classificationists were creating their taxonomic structures for pragmatic, rather than purely scientific, concerns; as a result, characteristics of utility were often debated in relation to scientific validity.[19] Nonetheless, the process of classification development in these fields was historically described as scientific and logical, with universal truth as its (perhaps unattainable) ideal.[20] Just as with biological taxonomy, however, those historical ideals have been discarded. Current scholarly consensus describes the classificatory enterprise as an argument or perspective, one that should align with accumulated evidence but that equally constitutes one of many interpretations of that evidence (for instance, see Hjørland and Albrechtsen 1995; Beghtol 2001). Concurrently, there is a recognition that, if previous ideals of scientific neutrality are impossible to achieve, classificatory devices might usefully abandon all pretenses toward appearing "unbiased"—that an unbiased taxonomy is a fantasy (Olson 2002; Tennis 2012; Fox 2016).

When we release taxonomic structures from their historical associations with universality and neutrality, it becomes possible to see how the relatively transparent argumentation of taxonomic structures can work toward, rather than against, pluralistic conceptions of knowledge. Taxonomic constraint becomes an invitation to creativity, rather than the opposite. For instance, if our conception of taxonomies remains in the historical mode, it is easy to imagine that gender fluidity cannot be represented taxonomically. But this is the case only if we begin with Male and Female as the primary manifestations of gender identity. If, however, we understand taxonomies as contingent interpretations of evidence, we can use taxonomic structure as a mechanism to explore alternate configurations of gender categories, arranged according to different salient relationships, where Male and Female classes are subsidiary rather than central. To illustrate, here are two possibilities, each of which offers a slightly different perspective on such relationships:

GENDER TAXONOMY #1
```
Gender
    Strongly typed genders
        Male and female-associated genders
            Cis genders
                Female genders
                Male genders
            Trans genders
                Female genders
                Male genders
        Androgynous genders
    Loosely typed genders
        Fluid genders
        Static genders
            Singular genders
            Plural genders
```

GENDER TAXONOMY #2
```
Gender
    Open (boundary-crossing) genders
        Union genders (both male and female aspects)
            Fluid genders
            Static genders
        Alternate genders (neither male nor female in
        expression)
            Determinate genders
            Indeterminate genders
    Oppositional (boundary-maintaining) genders
        Male (cis and trans) genders
        Female (cis and trans) genders
```

Such alternatives, which relate gender categories according to variant organizing principles, can coexist with each other, demonstrating contrasting patterns of gender differentiation. Taxonomic structure is perfectly accommodating to such experimentation. If we are not used to thinking of taxonomies as vehicles for expressing plural or partial accounts of the world, it is due to lingering habits of historical design patterns rather than structural necessity.[21]

There can be, moreover, multiple options for generating data with taxonomies. Historically, some use cases required placement within a single taxonomic class: a physical library book cannot be on two shelves at once, and science has traditionally limited a single biological specimen to identification with a single species. But this is not an inherent restriction of taxonomic structures. We can assign multiple values to a single object if we want! Along these lines, even a traditional binary representation of gender, limited to Female and Male values, can be implemented such that a single person can be described with one value, two values, no values, or dynamic values (e.g., a single value that can be changed at will). These implementation choices can be made subsequent to the design of the taxonomic structure itself and can vary for each context of data creation—indeed, for each instance of data creation. In other words, although we have tended to see a taxonomy as a monolithic imposition of some perspective onto the world, the data generated with a taxonomy is actually the product of many hands via the aggregation of many independent decisions. At each step, creative choices are possible, even as previous decisions will likewise enact constraints. This use of taxonomies to create data can accordingly be considered an extension of the data design.[22]

It may seem a little strange to talk approvingly about the potential for creativity in the use of infrastructural elements, such as taxonomies and data schemas, that are often relied upon to ensure data standardization. Isn't the control of data values—limiting them to a set of defined choices—one of the primary reasons for using a taxonomy? Isn't creativity contrary to that goal?

Structural Constraint as an Invitation to Creativity

The thing is, we've never been able to eliminate creativity from data collection. Fundamentally, an endorsement of creative practice merely surfaces and validates the way that data creation already works. No set of data standards is employed in a robotic manner, and data is already a lot more various than we tend to acknowledge. The only shift here is the suggestion that we might exploit those judgments more purposefully and transparently, with an eye toward the value produced through creative action—creative action within the constraints of standardization, that is.

Without those constraints, creativity in data becomes more difficult to discern and appreciate, because the parameters of individual decisions are obscured. In the adventure essay for this chapter, I referred to the drawer that many of us have in the kitchen, the one full of odds and ends: scissors, spare keys, take-out menus for local restaurants, measuring tape, flashlights. The jumble drawer in the kitchen can be an effective organizing structure when it's a private enterprise managed by a single person. Even without an explicit articulation of the jumble drawer's structure, a single person can maintain a relatively coherent sense of what goes in that drawer and what doesn't—that the soy sauce and ketchup packets go in the cutlery drawer with the extra disposable chopsticks and not in the jumble drawer, for instance. But any jumble drawer tends to drift as time passes—wait, where did I put the Sharpie again?—and when a jumble drawer is shared, that process is accelerated. It's rare for a household to talk about what goes in the jumble drawer; we all tend to assume that people see that drawer in similar ways. But inevitably this is not actually the case, and the jumble drawer devolves into chaos.

In our house in North Carolina, for instance, my partner Jason and I have a systematically arranged tool cart in one of the upstairs closets. But Jason is constantly placing tools from the cart into our kitchen jumble drawer after he uses them downstairs, just in case he might use them downstairs again. Then I open the jumble drawer looking for the scissors and find it inexplicably full of mallets, hammers, and wrenches. Well, this is clearly wrong! The jumble drawer should only contain tools for measuring and labeling or opening mail and packages: not mallets, not screwdrivers, and not whatever else Jason happened to be using that day. Have I ever actually explained this to Jason or even to myself? Well, no; it just seems obvious that's what the jumble drawer should be. (Chapter 3 describes this resort to obviousness as the *duh* warrant.)

As you might predict, when it comes to the jumble drawer, Jason and I each think the other one is irrational. So we each correct the situation according to our private vision. Accordingly, the status of the drawer lurches unpredictably back and forth (he puts things into it and I take them out, and then we repeat the cycle). With one of us in control the drawer might make sense, to the extent that either of us is able to hold a structural vision relatively constant inside our own heads. With both of us, however, the drawer is a contested mess. Even as we try for balance, because we do want to live together in harmony, there aren't any constraints for us to fall back on (there is nothing to stop us from putting anything in or taking anything out), and although we mean well, we keep making decisions that the other can't fathom.

Online, uncontrolled user tagging, sometimes called *folksonomy*, is basically a jumble of data. The idea behind user tagging is that it's a good idea to get as many different positions on things as possible, so why not collect as many keywords as we can, contributed by anyone who wants to provide them? At its base this is a fine idea, but a real dialogue between perspectives needs rules for engagement, or else it's easy to have babble rather than discourse.[23] Without some kind of underlying structure to establish relationships between data values, the result is effectively a series of jumble drawers, as constituted by the independent activities of multiple actors behaving like many of me and Jason running around, sometimes in sync but sometimes at extreme cross-purposes.[24] I'm not trying to dismiss jumble drawers: we all have them. But there are also good reasons why we put most of our stuff into more structured environments, or why we make the bed every morning even when we're just going to mess it up again the next evening. This is especially true when we live with others and want to function in a caring, as well as efficient, environment.[25]

When I mention *others*, thoughts may turn to an other that has been conspicuously absent throughout most of this discussion: the ultimate user of a dataset or other information system (as opposed to the user of a taxonomy, who is typically a data creator rather than a data user). As the conclusion of this chapter's adventure essay observes, taxonomic structure may not align with anyone's "intuitive" sense of order.

Under the old, traditional view of taxonomies as the expression of universally true relationships, conflict with user needs, individually or collectively, would have seemed obviously irrelevant. If a taxonomy is describing the world as it really is, then it doesn't matter what people think or prefer. Under this orientation, even bibliographic classificationists, whose goals were relatively pragmatic, were generally indifferent to what users thought. Bibliographic classificationists were certainly aiming to develop systems that supported effective retrieval from library shelves—Ranganathan, as one example, positions the classificationist's goal as creating a helpful sequence—but the "help" was located in the depiction of coherent, transparent classificatory relationships that reflected the actual world, rather than some individual's potentially erroneous perception of it.

Even as current scholarship consistently describes the classificatory enterprise as the expression of arguments rather than the expression of truths, scholars have not embraced direct user input in the design process. Birger Hjørland's program of domain analysis, which takes pluralism of perspectives as one of its foundational tenets, nonetheless views user needs as an inadequate foundation for classificatory design decisions (Hjørland 2002; Hjørland

2004). In a wider environment in which user-centeredness has been generally celebrated, Hjørland's perspective can seem almost shocking in its apparent dismissal of user beliefs, values, and preferences.[26] Hjørland's underlying agenda, however, is to resituate rather than reject utility in the taxonomic context. Taxonomies, with their structural demands, derive their utility not as mirrors of private, individual opinion but as public articulations of argument. Even an idiosyncratic taxonomy, in other words, can't be a jumble drawer; it needs to be internally coherent, its logic traceable to a reader. The process of enacting a position within a taxonomy requires that it be examined and adjusted—indeed, artificially distorted—to render it externally interpretable. The utility of a taxonomy, then, is not about individual users finding items where they first look for them. The utility is in the public explanation of a system of relations where we can understand the basis upon which a thing has been described *this way* as opposed to *that way*. These explanations serve a social purpose rather than an individual one.

Although an individualistic rationale for utility has been predominant within the various disciplines associated with designing, building, and studying information technologies, this focus has come under recent critical scrutiny. Social goals, and the public institutions that support them, have languished in this environment—to our mutual detriment. As Ann Light, Irina Shklovski, and Alison Powell (2017) succinctly put it, given the immense challenges that we currently face as a global society, "ease is not serving us" as a rationale for information systems design.

Responding to this situation, Ryan Shaw (2019) advocates for a sense of duty, on the part of the information professions, to *explain* information in addition to *providing* it. Shaw acknowledges that these explanations and the institutions that provide them are always going to be partial and fragile and, in the words of Luc Boltanski, "more or less lousy." Still, given the alternatives, what else can you do?

I agree wholeheartedly. But you know what? Constantly reckoning with one's own limitations is hard. We need help—tools to keep us honest—or we will fail to be properly skeptical of ourselves and our own motives. This is, for me, one of the most important advantages of taxonomies. As elements of data infrastructure, they are partial, fragile, and more or less lousy. But they force a measure of honesty upon us. Taxonomies facilitate the goal of explanation, even as they can never in themselves complete it.

6

LABELS

This chapter looks at names: names that identify unique people, places, or things, and names that identify abstract classes and concepts. Sometimes, names seem to operate as arbitrary labels for whatever they represent: a name can be changed without changing the underlying data associated with it. Other times, names seem tightly integrated with the concepts that they label, and the name itself appears replete with data, so that changing the name would result in data loss. I take stock of these contradictions.

In the adventure essay, I tell stories about

- A mistaken name that I like.
- Greece and Macedonia.
- Mrs., Miss, and Ms.
- A magic wand.
- The Yellow Peril, Orientals, and the Chinese virus.

In the reflection essay, I ruminate over

- Common-sense distinctions between concepts and labels.
- Distinctions between concepts and labels in the practical literature of controlled vocabularies.
- Empirical realities of extracting concepts from documents.
- Reconfiguring concept-label relationships.

ADVENTURE: A MISTAKE ON MY MASTERCARD

A Mistaken Name That I Like

In November 2019, I began making arrangements for a research trip to Humboldt University in Berlin. I duly learned that, according to the University of Copenhagen's policies, all air travel had to be booked by Ida, the department's travel administrator, through the university's dedicated reservations system. Moreover, the purchase would require my corporate Mastercard. I didn't have a corporate Mastercard? I would have to obtain one.

It took a week to get the Mastercard application submitted, approved, and processed. Then when the card was finally issued, I was notified that it would be sent via post to my home address. This was disappointing news. Our residential mail service was erratic, and even domestic mail could take a week or longer to arrive. What if all the seats on my preferred flights were booked before I got the card? But the notification message concluded with instructions to access my new account on the Web. Hmm. If I could find the card number online, perhaps Ida could proceed with my itinerary.

I logged in and began to scan my account information. Name, address . . . would there be any way to find the card number? Wait, name . . . I blinked twice, and looked again. Then I laughed. Someone, somewhere had misspelled my name, but not in any of the normal ways (Melaine, Melany, Melonie, etc.). No, this time, instead of *Melanie* Feinberg, my new Mastercard was issued to *Meanie*. Meanie Feinberg!

In retrospect, it's easy to see how such an error could occur. Although the application was correct, someone along the way had left out the *l* when typing my name. Indeed, I was amazed that this misspelling had never occurred before. And yet, after an entire lifetime of having my name constantly misspelled, being called another name entirely, or simply being called no name at all—the result of having a twin sister with a similar but more common name—finally, a brilliant mistake!

I really, really wanted to have a Meanie Mastercard, but I was concerned about more delays with my airline tickets, and so I called the card issuer to get the problem fixed. Ultimately, I received a card with the correct name, although the letter that accompanied the card was mysteriously addressed to *Meanie*. So I do have official evidence of my brief time as Meanie Feinberg (see figure 6.1).

Why did I find this particular mistake so marvelous? The Meanie Mastercard tickled my fancy because it integrated, within a single mechanism, a private aspect of my *identity* with my public *identification*. Outwardly, I present as good-natured and unthreatening, rather than disagreeable and peppery. But the Meanie mistake recognized my inner reality: cutting, crotchety, and cursing like a sailor. In contrast, my actual name conveys nothing about my internal qualities, even as it disambiguates me from other people reasonably well. Effective identification, in other words, need not have anything to do with

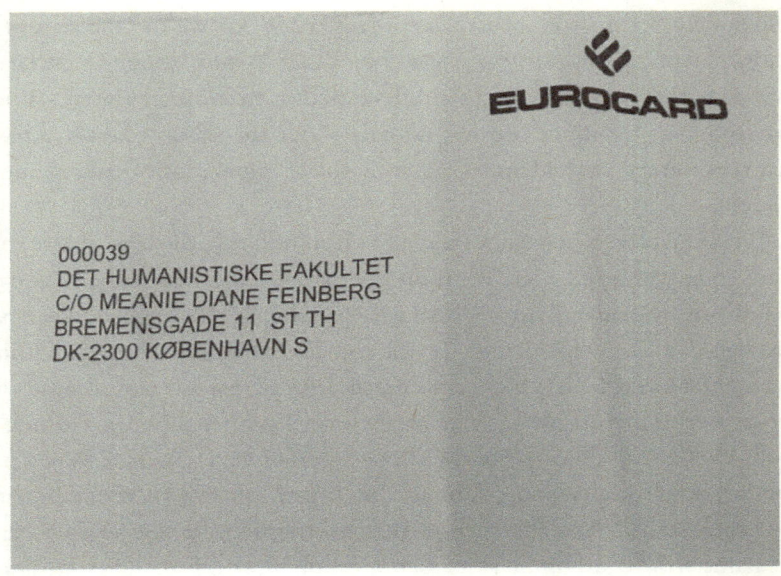

Figure 6.1
A letter addressed to *Meanie Feinberg*, which accompanied my new corporate Eurocard (the brand name for Mastercard in Europe).

identity. A label that operates solely as an identifier is devoid of additional data; a pure identifier merely designates an item without further describing it.

If we think about the purpose of names in daily life, we are likely to focus on the identification function first. We use names to distinguish between two separate, although potentially similar, entities, to tell *this* thing apart from *that* one. When we use names primarily as identification, it doesn't much matter what the name actually is: we can call our items A and B, or 137921 and 635792, or Batman and Robin. Anything works equally well, and the label itself is meaningless. For identification purposes, my name might be Melanie or Melinda or A or 137921 or Batman. It's just an arbitrary choice.

Sometimes, though, names do more than identify. The label that constitutes a name isn't arbitrary but incorporates its own layers of meaning, beyond the concept that the label is supposed to represent. If labels such as 137921 or Person B would merely identify me, Meanie does something more than that. Meanie has an association with an aspect of my inner essence, something that makes me a unique being. While my relationship with Meanie was a serendipitous accident, we sometimes choose names deliberately, in an attempt to assert identity. When we use names in this manner, naming an entity is a way of calling it into being, of declaring its existence and proclaiming its differentiation from other things. Where a meaningless identifier conveys no data about the thing it labels, a name that is entwined with identity can encapsulate quite a lot of data about its referent.

Consequently, when names serve to claim identity, the label that we use tends to matter quite a bit. We want the expression of the name to convey some of the named individual's distinctive nature. For that reason, when parents agonize over which name to pick for their children, the name's utility as a unique identifier is seldom considered. Instead, parents seek a name with certain expressive qualities, either for its formal aesthetics (what it sounds or looks like) or for what it evokes (what it signifies or represents). When parents write to social advice columnists in distress because their child's name has been "stolen" by someone else, the letter writers are not angry because the name will be less useful as an identification mechanism.[1] They are angry because the name's ability to convey identity appears diminished. If too many children are named Maya, then the name Maya's special connection to their child's identity seems compromised.

Arguments over stealing baby names can seem absurd to outsiders. However individuals might feel about certain names and what they signify, labels don't actually define identity. If a baby named Maya grows up and changes her name to Mildred, the woman formerly known as Maya remains the same person. Her identity doesn't change. Still, the fact that Mildred cared enough about her name to undertake the onerous process of changing it is indicative of the complicated feelings that many of us have about names, about identity, and about the relationship between names and identity. The people close to Mildred, also, may have complicated feelings about such a change. Some people may find it difficult to disassociate Mildred from her former name and be sad or even angry about it, seeing a change of name as a rejection of family history or cultural heritage. Indeed, the same people that think it's silly to complain about stealing baby names might feel regret about Mildred's change of moniker, objecting that she just seemed like a Maya, as if indeed the name did reflect some essential quality about the person using it. At the same time, others might use the same rationale to approve of Mildred's decision, imagining that Mildred is a better fit or is somehow more suitable, rather than just an arbitrary label and an arbitrary choice.

The Mildred example is a simple one, in that no particular rationale for changing a name was specified. Perhaps Mildred just never liked her birth name very much. But sometimes people change their names for reasons strongly connected to identity: to make a name seem like less of an arbitrary identifier and more expressive of who they are. Members of the Nation of Islam, such as Malcolm X and Mohammad Ali, changed their birth names to assert liberation from their forbears' enslavers. Trans and nonbinary people, likewise, may change their names to evoke their true genders or to avoid gendering entirely. When names are specifically chosen for reasons of identity, being called by the wrong name is not merely an identification problem but can be considered an act of violence.[2]

Greece and Macedonia

When it comes to identification and identity, names of places bear similarities to names of people. Many place names are shared without conflict. Within the United States, there are many cities and towns with the same name. I grew up in Long Beach, California; there is also a Long Beach in New York, New Jersey, Washington, and probably in other states as well. I don't ever remember anyone in my hometown being concerned in any way about other towns named *Long Beach*. Even when locations are geographically close and yet under different sovereignty, similar names are commonly treated with indifference. For instance, the country of Mexico coexists peacefully with the bordering US state of New Mexico. As with many such situations, the history is tangled: the region known as New Mexico has been part of both countries (first Mexico and then the United States). Meanwhile, of course, local communities have existed, arisen, adapted, and persisted as borders changed around them. Nonetheless, it is not currently a matter of public concern that New Mexicans, in keeping that name, are asserting anything in particular about the identity of their territory in relation to their neighbors across the border—that is, that bordering regions of Mexico should really be part of New Mexico (in other words, part of the United States), or vice versa, that New Mexico should really be a territory of Mexico once again. Should a citizen of Albuquerque begin agitating for the country of Mexico to change its name, it's hard to imagine many others joining the campaign.

But a similar situation in southeastern Europe has followed a different pattern.[3] Since antiquity, a region in the southern Balkans has been associated with the name *Macedonia*. Alexander the Great, conqueror of a legendary empire, originated from Macedonia. In Alexander's time, Macedonia was primarily populated by Greek tribes. But later, Slavic peoples also lived in Macedonia. Meanwhile, various empires ruled the area (Persians, Romans, Byzantines, Bulgarians, Serbians, and Ottomans). By the late twentieth century, places and peoples claiming Macedonian heritage were split among multiple countries. In particular, neighbors Greece and Yugoslavia both had provinces called Macedonia.

After Yugoslavia disintegrated in 1991, its former constituent republic declared independence as a sovereign nation, intending to retain *Macedonia* as its name. But this angered the neighboring Greeks, who saw it as an encroachment on their identity, in which the heritage of Macedonia, through Alexander, was distinctively Greek (as most of the people in the Greek province identified themselves) and not Slavic (as most of the people in the former Yugoslavian province identified themselves).

Within Greece, opposition to the new country's proposed name was so strong that, in 1992, one million people marched in the Greek city of Thessaloniki (capital of the Greek province of Macedonia) to protest the use of the name Macedonia by their neighbors. (As documented by Eurostat in 2019, the entire population of Greece was 10.7 million people.) Seeking a compromise, the new country agreed in 1995 to temporarily rename itself The Former Yugoslav Republic of Macedonia while negotiations continued. Under this temporary solution, the new country was admitted to the UN, and its sovereignty was recognized by many UN member states. But Greece never supported this compromise, and the situation remained at a stalemate for more than 20 years. During this period, Greece referred to its neighbor only by the acronym Fyrom, to avoid legitimizing the association of the Former Republic with the name Macedonia. Meanwhile, Greece blocked the Former Republic's entry into NATO and the EU. In response, to assert their rightful association with the name of Macedonia, the Former Republic incorporated references to Alexander the Great throughout Skopje, the capital (the airport was named for Alexander, a huge statue was installed, and so on). Finally, in 2019, the two sides agreed on *North Macedonia* as an acceptable compromise. The country changed its name, and Greece removed its objections. North Macedonia finally joined NATO in 2020. Its quest for EU membership, however, continued. In 2020, Bulgaria demanded that North Macedonia describe its language and history as

Bulgarian in origin rather than *Macedonian*, further complicating the process (Georgevski 2020).

I became aware of the Macedonia squabble in 2018, as the two nations began again to negotiate and the global press took notice (once again there were mass protests in Greece). In the pages of the daily *New York Times*, these stories took on a bemused air. What a strange conflict, the reporters implied, with their dueling claims on the patrimony of an ancient emperor. But while the American accounts had a wry aspect, the implications, which imperiled the continued integration of the Balkans into greater Europe, were decidedly not trivial. And even in 2018, after a stalemate of 26 years, one hundred thousand people marched in Greece in opposition to the name of their neighbor. That is a huge protest, especially for a long-stagnant situation. This was clearly a significant issue for a good number of people.

The scale of the Macedonia dispute fascinated me: incredibly intense and yet intensely local. Away from the southern Balkans, the elements of identity that seem so vitally important to the Greeks and Macedonians and that seem so inextricably bound up in the name of Macedonia become difficult for outsiders like me to comprehend. After all, if we are concerned merely with effective identification, there are other, more confusing situations globally: for instance, the neighboring African countries called Democratic Republic of the Congo and Republic of the Congo, which are both named for the same geographic area, the basin around the Congo River. Here, too, the history is complicated. Before 1997, the DRC was called Zaire, but before that, it was the Belgian Congo, and then briefly the Republic of the Congo. The current Republic of the Congo was previously part of Equatorial Africa, a French colony. In the passing of political control from group to group, and in the mix of tribal affiliations sharing the same region, the situation in the greater Congo area is not dissimilar to the situation in the greater Macedonia area. But in the Congo, the naming challenges seem limited to problems with identification rather than identity, and identification difficulties can be managed. When necessary, the two Congos can be disambiguated by various means: by using their full names (the Democratic Republic vs. the Republic) or by referring to their capital cities (Congo-Kinshasa or Congo-Brazzaville). We're all familiar with these kinds of strategies for sharing names but keeping concepts separate. Families adopt similar mechanisms, without drama, when children are named after parents and grandparents (which tends to be seen as respectful homage rather than base thievery). The Bush family, famous in American politics, includes three George Bushes across three generations, differentiated by their middle initials: George H. W. Bush, George W. Bush, George P. Bush. Two of the Georges were president of the United States within ten years. And yet the world coped perfectly well with this potentially confusing situation.[4] Why couldn't Greece and Macedonia do the same?

Or so it seemed to me, at my breakfast table in Durham, North Carolina. I could read about the anger that fueled mass protests in Thessaloniki, but I couldn't feel it in the same way as the Greeks and Macedonians. It is this inability to viscerally *get* the significance of the Macedonia situation that encapsulates the dilemma of names. That names can do more than identify, on this there is no doubt. Otherwise it would make no sense for me to exult in the Meanie Mastercard, or for Macedonians and Greeks to persist in conflict for 27 years. But although all names have some identification function, only some names evoke feelings of identity, for some people, in some places, at some times, and with different degrees of strength. When I receive a Durham County property tax bill addressed to *Melaine*, I feel a passing annoyance, but then I just ignore the error. And yet the Meanie Mastercard makes me smile again just thinking about it. The name of Macedonia causes an extended international incident; the name of the Democratic Republic of the Congo doesn't. When is a name itself replete with significance, and when is a name just an arbitrary label that might just as well be replaced by some other label? It's unpredictable. It's also unpredictably important. A name that's terribly significant to one person, so fraught with feelings of identity in a way that seems so very obvious, may be completely meaningless to another person, who sees it as mere identification. But we all have a Meanie, even as we all find it difficult to understand a Macedonia.

Mrs., Miss, and Ms.

I'd like to think that the preceding paragraph demonstrates a calm, thoughtful rationality regarding the status of names as data, and the relationship between identity and identification at work there. But when a name is really significant to a person—when we feel the emotions of Greeks and Macedonians deep within our guts—it is hard to be detached, especially if we feel that a name is being used to suppress our identity rather than to express it. For me this happens when the occasional undergraduate calls me

Mrs. Feinberg. Rationally, I know that the student simply has no idea that addressing me this way has suffused me with incandescent rage. Indeed, the student is undoubtedly attempting to be polite. Still, my anger burns with the fire of a million suns.

Why can't I treat *this* name as an arbitrary identification mechanism? It's not that I'm tetchy about titles: faculty in my field tend to use first names, and I encourage students to call me Melanie if they are comfortable with that. But Mrs.—no. And it's not because *Mrs.* is incorrect, although it *is* incorrect (I'm not married, but I have a long-time partner and I wear a ring, so it's easy to make that assumption). If I heard a student addressing one of my married colleagues as Mrs. Hughes-Hassell or Mrs. Haas I'd get pretty steamed also.

Why the fury? One answer is that *Mrs.* Feinberg, in contrast to *Dr.* or *Ms.* Feinberg, invokes a society in which the marital status of women matters but the marital status of men does not. So if *Meanie* reveals some essential truth about me in a satisfying way (it expresses and validates my identity), *Mrs. Feinberg* contradicts an essential truth about me in a distressing way (it suppresses and disrespects my identity). This is a personal reason having to do with my individual values and how I see myself.

But there is another reason also. As it stands in 2020, *Ms.* rather than *Mrs.* has been the professional standard for a generation. As such it expresses *social* acceptance of gender equality—in some small degree, anyway. In other words, it's not just that *Mrs. Feinberg*, as a form of name, contradicts my own preferences. Personally, I'd just as soon abandon titles altogether, but it doesn't drive me crazy when someone calls me *Dr. Feinberg*. *Mrs. Feinberg*, in contrast, undermines the social acceptance of gender equality that *Ms.* represents, exposing its fragility. When someone calls me *Mrs. Feinberg*, they are telling me that gender equality in our society is illusory, or at least vulnerable.

It's common, actually, for names to be associated with social privileges and legal rights. In particular, when one group is systematically denied the rights and privileges associated with a certain name, being acknowledged with that name takes on special importance. As late as the 1960s, White Americans, particularly in southern states, would routinely address Black Americans informally, in a manner that would be considered rude or offensive if used in addressing White Americans. During this period, the standard convention for addressing women was to use *Miss* or *Mrs.*—*Ms.* had yet to become popularized. So when the Black civil rights activist Mary Hamilton was arrested in Alabama in 1963 for protesting against racial segregation, White officials purposefully demeaned her by calling her *Mary* and not *Miss Hamilton*. In court, Mary Hamilton declined to respond to the White judge until she was addressed as *Miss* Hamilton—until she was named as a White woman would be named and symbolically accorded the same rights and respect. The judge refused to address Mary Hamilton as *Miss*, and he sentenced her to jail for contempt of court (Domonoske 2017b). Mary Hamilton sued—and won. As established by the US Supreme Court, all witnesses in a court of law must be addressed using the same form of name.[5]

This case vividly illustrates the density of data that a name may hold. Just like my earlier example of Maya, who changes her name to Mildred, Mary Hamilton remains the same, whether she is called Mary or Miss Hamilton or anything at all. In my hypothetical example of Maya and Mildred, people in Mildred's life might have personal feelings about her name change—they might approve or disapprove, be happy or sad. But Mildred is not treated differently from Maya in any systematic way. In the world of hypothetical examples, in which everything operates with consistent logic, exchanging one label for another (Maya to Mildred) makes no difference. But the real world does not operate with consistent logic, and labels aren't equally random strings. In the real world of 1963 Alabama, *Mary* interacts with the legal system differently than *Miss Hamilton* would. Everyone acknowledges this. That's why Mary Hamilton demands to be addressed as Miss; that's why the Alabama judge refuses to do so; and that's why the US Supreme Court ultimately rules in Mary Hamilton's favor. If the legal rights and social privileges associated with the dominant group (White people) were not encapsulated in the name *Miss Hamilton*—if one name really could be replaced by another without changing what the name represents—then none of this would matter. But of course it matters. And even more importantly, the characters in this drama all know that the data in *Mary* is completely different from the data in *Miss Hamilton*. There's no confusion about that at all.

In a flourish of history, Mary Hamilton, who sued to establish her right to be addressed as Miss Hamilton, had a roommate, Sheila Michaels (Kay 2007; Domonoske 2017a). Sheila Michaels was also an activist, but she was White. Sheila Michaels became a fighter for the use of *Ms.* to replace the use of *Miss* and *Mrs*. In fact, Sheila Michaels was introduced to the term *Ms.* by seeing a piece of mail addressed to her roommate, Ms. Mary Hamilton. Intrigued by the notion of an honorific that did not disclose a woman's marital status, Sheila began to take up that cause. But as she recalled later, Mary Hamilton was not interested in changing the system of honorifics. As a Black woman, Mary Hamilton wanted access to the current system of naming in order to assert her equality with White women. Gender equality was secondary to her. As a White woman, however, Sheila Michaels already had the privileges that Mary Hamilton sought. In 1963, the data rolled up in Sheila Michaels's name was fundamentally different from the data rolled up in Mary Hamilton's. If one rude judge in Alabama happened to call Sheila Michaels *Sheila* rather than *Miss Michaels*, it might not mean anything at all, in terms of Sheila Michaels's ultimate treatment by the courts. In Mary Hamilton's case, being called *Mary* was a systematic mechanism that marked her as a member of a marginalized social group (Black Americans) and accordingly as receiving a different set of rights from the dominant social group (White Americans). But in Sheila's case, her name didn't work that way. For Sheila, being addressed informally by a single judge would be an aberration. It would just be noise, something to clean out of a dataset. In contrast, cleaning up Mary's data would misrepresent Mary's experience. As names, *Sheila* and *Miss Michaels* had less data (and were more synonymous) than *Mary* and *Miss Hamilton*. In the context of the United States in the 1960s, Sheila's status as a White person meant that she, unlike Mary, had the freedom to pursue a different kind of naming action: to replace *Miss* with something else, a new type of name that would represent different data. Both Mary's naming action (changing the use of an existing label) and Sheila's naming action (creating a new type of name that expresses different data) were meant to intervene in the world—to enact social change—and not just to describe the world as it currently exists.

A Magic Wand

So far, I've talked only about names that directly refer to people and places: proper nouns. But there is a sense in which labels for concepts such as *pink* or *table* are equally arbitrary. Just like Maya could change her name to Mildred and yet retain the same identity, we could change *pink* to *light red* (or to JR351 or anything else) and maintain that the concept continued to exist as it always had. As a thought experiment, let's imagine doing just that. I'll wave my magic wand, and . . . switched! Everything that was previously labeled *pink* is now *light red*.[6] The cherry tree outside my window is blooming with beautiful light red flowers right now. Light red has never been my favorite color, but the cherry tree is pretty against the blue sky. What about you? Do you like light red?

If that question is surprisingly difficult to answer, the reason is that all labels have traditions of usage, and sometimes that history infuses the label with a lot of additional data. For me *pink* is associated with gendered stereotypes of femininity, and I've always shunned it. I simply can't separate *pink* from being told to sit still and act like a little lady. If the name of *pink* changed to *light red* even as the range of hues associated with that concept stayed the same, would I still hate it? I'm really not sure. Logically, *light red* would still be the color of bubblegum and Disney princesses. And yet so much of the way that I experience this color is bound into the word *pink* that *light red* really seems different. My inability to answer this question easily—would I hate *light red* just as much as I hate *pink?*—demonstrates how labels like *pink* have their own data in addition to the data of the concepts with which they are associated.

The *pink* case is my little fantasy. But it underscores the pervasiveness of name data. Apparently innocuous labels can contain vast amounts of historical data, data that is deployed as evidence to make decisions in the world, such as a male tendency to disdain pink, and the associated reification of gender stereotypes. Does that mean then that we should all take out our magic wands and commence to deploy them? As my previous examples show, name changes can be fraught with conflict. Whether Macedonia or *pink*, one contingent will mourn a loss of name data as another rejoices in its banishment, while a third group finds itself struggling to understand what all the fuss is about. But although it's never easy to select between competing interests, some cases do seem easier than others. The names that we use for groups of people—often, to designate race and ethnicity—are particularly salient here.

Orientals, the Yellow Peril, and the Chinese Virus

In 2016, legislation initiated by Representative Grace Meng removed the word *Oriental* from Title 42 of the US Code, replacing it with the word *Asian* (United States, Office of Congresswoman Grace Meng 2016). In a press release, Representative Meng noted that "Many Americans may not be aware that the word 'Oriental' is derogatory. But it is an insulting term that needed to be removed from the books."

Beginning in the mid-1990s, my father and I had many discussions about the word *Oriental*. I told him that he shouldn't use it, but he didn't understand what had changed. Indeed, when I was growing up in the 1970s, the term *Oriental* was widely used. (The text of the laws that Meng's legislation changed was written in the 1970s also.) Its meaning, as defined in my dad's 1960s-era college dictionary, was geographical: it referred to the East, to Asia. My father, as one of the many Americans referred to in Meng's press release, did not find anything wrong with this dictionary definition. As late as 1993, the fourteenth edition of *The Chicago Manual of Style* included *Oriental* in a list of examples demonstrating proper capitalization for names of ethnic and national groups. But just a few years later, the *Chicago Manual* examples had been quietly revised. *Oriental* had been replaced by *Asian*.[7] What had changed? In one sense, nothing. The term *Oriental* hadn't gained a new, disrespectful meaning. But *Oriental* had always been connected to the racist history of the Yellow Peril, and more people had started to recognize that.[8]

In the late nineteenth and early twentieth centuries, the phrase *Yellow Peril* was used in the United States (and in the West generally) to describe a fear of aggression by Asian nations. Abroad, this idea was used to justify colonial expansion in Asia. Internally, it was used to promote restrictions against Asian immigrants. During this period in the United States, *yellow* might also have been used in general discourse to describe people from Asian countries or of Asian descent, and this usage would have been unremarked and ordinary. But the association between *yellow* and *Yellow Peril* was not subtle—*yellow* marked people of Asian descent as foreign, no matter their country of birth or citizenship.

Oriental might have been normal in my childhood, but *yellow* was definitely not. I didn't learn about the Yellow Peril until high school, but *yellow* was clearly unacceptable. *The Chicago Manual of Style* didn't start including capitalization examples for nationalities and ethnic groups until the 1969 twelfth edition, but *Oriental* was included there also and not, pointedly, *yellow*—although color terms such as *black* and *white* were used as examples of terms to be lowercased.

So how was *Oriental* connected to the Yellow Peril? An instructive example arises in the writing of Sidney Gulick, an American minister who lived in Japan for 25 years and sought to promote sympathy for Asian immigrants in the United States. In the first chapter of his book *The American Japanese Problem: A Study on the Racial Relations of the East and West* (1914), Gulick set out his goals: to help "the white man" understand "the yellow man." After that first chapter, however, Gulick completely avoids the term *yellow*, using *Asiatics* or *Orientals* instead. With this rhetorical move, Gulick seemed to propose that *yellow* should no longer be used, and that his alternate terms are more acceptable—more open-minded, as he encouraged his readers to be.

Nonetheless, a paradox emerges. Gulick's argument for tolerance was based on the capacity for Japanese immigrants to be fully assimilated—to lose their Japanese identity entirely. (He offered as evidence cases of Japanese children taken in and raised by Americans, whose connection to Japanese culture had been severed. To Gulick this was positive.) In promoting this kind of assimilation, Gulick, despite his avowed goodwill, reinforced a key element of the Yellow Peril—that immigrants from Asian countries who retain their heritage are not really American. *Orientals*, in Gulick's book, were still set apart. In this way, even as *Orientals* might have seemed less pejorative than *yellow*, it still performed many of the same functions as that earlier word. In fact, it could be argued that *Orientals* was actually a more harmful term, because it appeared to be innocuously geographic—like *European*—even as its usage went beyond mere geography.

I wish that I could say that I explained all this well to my father, but I was never able to. Our discussions always became arguments about who was right and who was wrong: what *Oriental* really *meant*. But in situations like that of *Oriental*, outcomes are paramount. Names become important as data not from being used correctly or incorrectly but from the accumulation of their use and the aggregation of those usages over time. Those accumulations can lead to harm—sometimes devastating harm. And intentions don't matter. I might imagine, for instance, that the elderly Sidney Gulick was horrified when the US government imprisoned Japanese Americans in camps during World War II, out of fear that even third-generation citizens would have a stronger connection to Japan than to the country of their birth. But even as Sidney Gulick's 1914 book exhorted Americans to welcome Japanese immigrants in California, it likewise, partly through the names that it used, made their subsequent internment possible. What I should have said to my father was this: even as names are generally tricky and confusing, situations regarding the names of races and ethnicities are easy in one respect. We should always scrutinize their potential for harm.

I wrote the first draft of this chapter from the ugly and uncomfortable sofa in our Copenhagen rental apartment as I worked from home, a result of the COVID-19 lockdown in Denmark and around the world. COVID-19 is the official name for the disease caused by the coronavirus discovered in 2019 (subsequently to cause a worldwide pandemic and global chaos in 2020). In selecting a name for COVID-19, the World Health Organization (WHO) followed its 2015 guidelines for naming new diseases (World Health Organization 2015), which aim to avoid mention of peoples or places in disease names. Why? Because no place or people wants to be linked to a disease, even if the association is oblique. The 2003 coronavirus disease SARS (severe acute respiratory syndrome) does not appear to reference a locality, but Hong Kong is a SAR (special administrative region) in China, and Hong Kong saw many cases of SARS. To Hong Kongers, the name SARS carried an implication that Hong Kong was responsible for the outbreak (Branswell 2020).

Are the WHO's 2015 naming guidelines overly sensitive and unnecessary? The case of COVID-19 suggests not. In the United States, President Trump made persistent attempts to use the name *Chinese virus*, claiming that this term was "not racist" but was merely referring in a neutral and objective way to the location where the virus was first identified. Trump allies joined him in referring to the Chinese virus. Meanwhile, Asian Americans across the United States reported a sharp rise in racially motivated taunts and threats. To make matters worse, even when President Trump tweeted his support of the Asian American community, his language reinforced a conception of Asians as more foreign and less American than other immigrants. The virus "is NOT their fault," Trump tweeted, "They are working closely with us to get rid of it. WE WILL PREVAIL TOGETHER!" The implication is clear. Asian Americans are not "us." They have a stronger relationship to the "Chinese virus" than "we" do.[9]

When I received notification about my Meanie Mastercard, I thought it was awesome. Someone else, however, might have reacted quite differently; they might have perceived it as a terrible insult or humiliation. Similarly, some people are enraged when their baby's name is "stolen," while others don't mind how many people share their child's name. Although people can perceive these situations very differently, the ultimate consequences of such cases are minor and limited to individuals. If someone were to mistakenly believe that my name was actually Meanie, it might cause momentary confusion or embarrassment, but that would be all. If multiple children in one family have the same name, the effects on the children's lives are typically slight and manageable (recall the George Bushes). Even in a situation like

Greece and Macedonia, where emotions are intense for millions of Greeks and Macedonians, the history of the Greeks as a people isn't actually diminished if a neighboring country takes the name Macedonia. It may seem awful to Greeks if Macedonia is so named, but nothing materially changes for Greece as a result.

But sometimes, using one name instead of another has more significant consequences. Calling me Mrs. Feinberg in my workplace signals that my professional status is not as important as my marital status—should staff reductions ever be necessary, for instance, *Mrs.* Feinberg is more vulnerable than *Dr.* Feinberg. When Mary Hamilton can be addressed as *Mary* in a courtroom, emphasizing her Blackness, we can anticipate that she is more likely to be convicted and will receive a longer sentence than *Miss Hamilton* would. And when as Asian person is referred to as *Oriental*, we should not be terribly surprised if that person's status as an equal citizen is questioned. There is historical evidence for all these assertions: women have been systematically fired because of their marital status, Black Americans typically receive harsher prison sentences than White Americans, and Asian immigrants' loyalties have been disputed. In all these situations, name data contributes to systemic harm, not just for isolated individuals but for entire groups of people. The only person affected by a mistake on my Mastercard is me. When Mary Hamilton was called *Mary*, rather than *Miss Hamilton*, this mode of address—and the discriminatory treatment that accompanied it—was perpetuated not only against her but against all Black Americans. The name was part of a coordinated system of exclusionary behaviors that identified a certain group, separated them from others, and dictated rights and privileges that were not accessible to them. As we well know and as examples like *Oriental* illustrate, these exclusionary behavior patterns have had a long historical reach. President Trump's crude and obvious use of *Chinese virus* illustrates this with sparkling clarity. Hotheaded Americans responded to this provocation by attacking and taunting fellow citizens and legal residents of Asian heritage, continuing the legacy of Orientalism. Through the use of one name rather than another, systemic harm was visited upon a historically marginalized population.

As I began this chapter in the spring of 2020, the coronavirus went from being a vague, faraway idea to a frightening and chaotic reality for the entire world. Everywhere, people struggled to comprehend a rapidly changing situation. What should we be doing, and what should we not be doing? In Denmark, as we attempted to adjust to a national lockdown, Queen Margarethe appeared on television with a special address to the nation. The Danish monarch gives a short televised speech every New Year's Eve, but spontaneous royal speeches are exceedingly unusual (the preceding one was just after World War I). When the queen's intentions were announced, the Danish media was agog. What would she say?

The queen reduced the circumstances around the pandemic to an essential simplicity. She began with an imperative: everyone must take this situation seriously, because it is especially dangerous for the most vulnerable among us. Fortunately, the queen reminded us, our duty here is straightforward. (Wash your hands. Hold your distance. Avoid physical contact. Stay home.) And yet, the queen continued, some people seemed to think that because their personal risk might be low, they could ignore these four simple rules—that they might, for instance, have parties and celebrate birthdays. Those people, admonished the queen, should be ashamed of themselves.

When the queen mentioned birthday parties, I smiled ruefully. No one wants to cancel a birthday party, including the queen, whose eightieth birthday celebration had been scheduled for mid-April. I, too, had been slow to cancel an extravagant trip that my sister and I had planned for our fiftieth birthday that August. Maybe it wouldn't be too risky to travel in the summer, I had been thinking. I heard the queen, and I was ashamed. How easy it was for me to think that my own actions wouldn't matter! How trifling I was.

Most viruses, of course, do not require coordinated intervention. Most names don't, either. But some names do. Our goal with names should be, as with stopping the spread of a pandemic, to avoid perpetuating harm toward the most vulnerable among us. When we narrow our focus accordingly, then the situation with names becomes straightforward and direct. Are we invoking, explicitly or implicitly, a group of people who have been systematically disadvantaged? Then we should take special care to ensure that the names we use do not inflict further injury. If we think that our individual actions don't matter—that we're just going about our daily lives, maybe—then we should think about having a birthday party in a pandemic. And we should be ashamed.

REFLECTION: WHAT'S IN A NAME?

Common-Sense Distinctions between Concepts and Labels

People can't collect, manage, or understand data without names. Even quantitative data is a count of something: some concept that people understand via a name, like Temperature or Height or Number of Clicks or Price. Non-count data, moreover, is often expressed as a named subtype or instance of the concept (or attribute) of interest. A patient visiting a doctor's office may be asked to list symptoms, which are then recorded in a health record: each data value (Headache, Fever) is a subtype of the attribute (Symptom). A good deal of data, therefore, involves the intersection between two names, or labels. As another example, the Variety (an attribute label) of an apple is Cox's Orange Pippin (a value label). As human beings who communicate in language, we find it difficult to comprehend either attributes or values without such names.

Even as we require names to make sense of data, it seems equally clear that the data is not reducible to the name. Indeed, when I say that *Variety* is the *label* (or name) for a data attribute, and that *Cox's Orange Pippin* is the *label* for a data value, I am saying that the attribute and the data value are more than the character strings *V-a-r-i-e-t-y* and *C-o-x-'-s-O-r-a-n-g-e-P-i-p-p-i-n*. These character strings serve merely as identifiers for the concepts that make up the real attribute and the real value. The real attribute identified by the label *Variety* is the idea of a subtype of fruit, with distinct characteristics (such as shape, size, color, texture, taste, and smell) as produced by a particular cultivar of apple. (A cultivar is a variety, or subset, of apple plants that has been purposefully bred to encourage certain characteristics.) Similarly, the real data value identified by the label *Cox's Orange Pippin* is the idea of the fruit of a certain apple cultivar. If we understand the attribute in this way, as the concept of a variety rather than the character string V-a-r-i-e-t-y, then its label is fungible. *Variety* can be replaced without changing the concept—without changing the data. We could relabel the attribute as *Cultivar* or *Type* or *Subtype*, and the data is not changed. We could relabel the attribute as *H89.03* or as an apple emoji, and the data is not changed. If, sometimes, we conflate data and label, it's probably because we have been using the label as an encapsulated description of the concept that it identifies. The label *Cox's Orange Pippin* can serve as a shorthand definition of the apple variety that we mean. It is not uncommon to do this, but it can be seen as sloppy data practice. *Variety* probably indicates this better. If we mean here the fruit of a distinct cultivar, *Variety* is accurate but underdetermined, potentially easy to misconstrue, and thus easy to apply inconsistently.

When I explain it, this distinction between label and concept probably seems quite obvious, even if you've never considered it before. When we are seeking apples of a particular variety, we are seeking the characteristics associated with the concept of a Cox's Orange Pippin—aromatic, juicy, tangy—and not just apples that have been associated with a particular set of letters. If there are apples of that variety that have been labeled in different ways (perhaps just *Cox's* or just *Orange Pippin*) that should not matter. Indeed, it could be a big problem if we do not recognize that multiple labels (Cox's, Cox's Orange Pippin) can equally identify a single concept. From this perspective, aggregating synonymous labels is a fundamental component of data management.

In everyday life, there are many familiar techniques to establish and manage the relationship between labels and concepts: dictionaries, thesauri, and gazetteers all perform this function. Controlled vocabularies created to facilitate information management work according to similar principles. In its simplest form, a controlled vocabulary is just a list of allowable data values. Using such a list for apple varieties would ensure that only a single label (for instance, *Cox's Orange Pippin*) is used in a dataset. A more complex kind of controlled vocabulary establishes the preferred label for a concept (*Cox's Orange Pippin*, for instance) and additionally notes any equivalent labels (*Cox's* or *Orange Pippin*). Such devices are long established. A library patron using a card catalog generations ago to find material about Orange Pippins under the letter O might have found a card that read "Orange Pippins. See Cox's Orange Pippins" and thereby been directed from an unpreferred synonym (Orange Pippins) to the preferred label (Cox's Orange Pippins). Similar mechanisms quietly facilitate information retrieval today. We tend not to notice if such devices are present in the search engines that we use, because redirection is automatic rather than manual. But they are often present, particularly for systems oriented around specific domains. For instance, when an information seeker queries the PubMed database of medical literature for *concussion*, the retrieval system automatically maps this synonym to the preferred label *brain concussion*, as defined in the controlled vocabulary called Medical Subject Headings (MeSH). We retrieve articles indexed with *brain concussion* even when that is not exactly what we asked for.

Distinctions between Concepts and Labels in the Practical Literature of Controlled Vocabularies

As a kind of data structure, the controlled vocabulary became formalized through the professional practice of indexing, that is, describing the aboutness (subject matter) of written documents. The people at the National Library of Medicine who use MeSH to describe the subject matter of medical literature are indexers.[10] In corporate environments, *indexer* has given way to fancier appellations. One of my first students at the University of Washington became a *taxonomist* for the Nordstrom department store Web site, managing relationships between synonyms like *dress* and *frock* and *gown*. Despite the upswing in glamorous job titles, fundamental processes and tools of indexing have changed remarkably little in the past 50 years. In 1960, the information scientist Brian Vickery (1960b) wrote of a new kind of controlled vocabulary called a *thesaurus*—different from the writer's source of synonymous terms, an indexing thesaurus is the most complex form of controlled vocabulary, incorporating hierarchical and associative relationships between its constituent concepts, as well as equivalence relationships between labels. By 1974, Dagobert Soergel produced a five-hundred page tome that outlined in meticulous detail how a thesaurus should be properly constructed and maintained. In 2006, when I was the teaching assistant for the future Nordstrom taxonomist, we still used Dagobert Soergel's book as a primary resource, updating little of his material.

The relative stability in the literature of controlled vocabulary creation is interesting because in some respects it is riddled with internal contradictions. The current National Information Standards Organization (NISO) standard for controlled vocabularies, released in 2005 (with minor corrections in 2010), demonstrates this situation well. The 2005 NISO standard both mandates a clear separation between concepts and labels and, at the same time, focuses almost entirely on labels. Indeed, concepts are almost totally absent from the standard—as they are almost totally absent from Soergel's book, and from other technical manuals and guidebooks for controlled vocabulary development.[11]

Instead of concepts, the literature of controlled vocabularies, as exemplified by the NISO standard, focuses on *terms*. Traditionally, *terms* are the basic unit of a controlled vocabulary. Indexers assign *preferred terms* to documents to express their aboutness. Each preferred term (sometimes called a *descriptor*) may be related to *unpreferred terms*, or synonymous labels. An example of this structure might be as follows:

```
Variety
    UF Cultivar
    UF Subtype
    UF Type
```

Variety is the preferred term, which the indexer uses to assign aboutness data to documents. Cultivar, Subtype, and Type are Variety's unpreferred synonyms. UF stands for *use for*. In other words, don't use the terms Cultivar, Subtype, or Type; use Variety instead. Unpreferred terms appear in the controlled vocabulary as follows:

```
Cultivar
    USE Variety
```

This indicates that documents should not be indexed with Cultivar but instead with Variety.

So, what *are* terms? Are they labels? To consider this question, let's look more closely at the USE/UF cross-reference, which indicates an equivalence relationship. The 2005 NISO standard defines equivalence relationships this way:

> When the same concept can be expressed by two or more terms, one of these is selected as the preferred term. The relationship between preferred and nonpreferred terms is an equivalence relationship in which each term is regarded as referring to the same concept. The preferred term in effect substitutes for other terms expressing equivalent or nearly equivalent concepts. (NISO 2005, 43)

Aha! This passage reveals that from the perspective of the NISO standard,

- Terms represent (express, refer to) concepts.
- When two terms represent the same concept, either might be selected as the preferred term.

From this discussion of the equivalence relationship, it appears that terms are indeed labels: inessential accessories to concepts.

If this characterization of terms seems straightforward, however, it is also rather strange. On the one hand, controlled vocabularies regulate labels for concepts. These labels have no necessary relationship to the concept and are replaceable at will. On the other hand, concepts (the foundation of the whole enterprise) exist only via the terms that instantiate them (arbitrary labels that can be added, removed, or changed for any reason). If concepts are what matter, and terms are inessential labels that can be substituted for each other at the complete discretion of the vocabulary creator, then why do controlled vocabularies focus on terms?

There is a simple answer and a complicated answer. The simple answer invokes legacy technology. What appears to be ambivalence about terms results from historical implementation constraints rather than muddled thinking. For many years, the technically feasible way to construct a controlled vocabulary was to create a list of labels (terms, preferred and unpreferred) and designate one label (the preferred term) as a proxy for the concept. Accordingly, in the literature of indexing, *term* became a kind of shorthand for "the way that we are currently representing a concept." In other words, the apparent focus on labels is misleading. The emphasis has always been on concepts. Vocabulary creators have merely lacked a reliable mechanism to identify and define concepts separately from the labels that represented them. Accordingly, concepts were associated with the currently preferred label. Terms, then, have a double identity: as concept identifiers (preferred terms) and as labels (unpreferred terms).

This simple answer implies that, if the technical infrastructure were available to identify, define, and relate concepts and labels separately, the notion of a *term* would no longer be necessary. There is evidence to support such a conclusion. One domain of evidence involves the literature of classification, of which the canonical examples are library classifications, such as the Library of Congress Classification and the Dewey Decimal Classification. To a great degree, classification and indexing are similar. They are both mechanisms for describing the subject matter of (generating aboutness data for) written documents. However, where indexing focuses on equivalence relationships, classification focuses on hierarchical relationships. The traditional function of a classification is to arrange documents in a helpful order (for instance, to arrange books on shelves in a way that facilitates discovery through browsing), while the traditional function of a controlled vocabulary is to improve retrieval (that is, to maximize the success of search queries). This distinction may seem abstruse, but what it means is that the practice of classification has focused on the definition and arrangement of concepts; classification has never been concerned with labels. The idea of *terms* does not exist in the literature of classification. Instead, the classes (categories) that make up a classification are identified by means of a notation, that is, a code—the notation shows the location of the class within the classification's hierarchical structure. Each class typically also has a label that describes it, but the label is clearly ancillary. In other words, whereas a controlled vocabulary identifies its constituent concepts by means of the currently preferred label (the preferred term), a classification identifies its constituent concepts by means of coded identifiers (a notation).

The traditional domains of classification and indexing collide and merge in more complex controlled vocabularies such as the thesaurus, which incorporate both hierarchical relationships (between *concepts*) and equivalence relationships (between the *labels* that refer to concepts). In many respects, a *faceted classification* and a *thesaurus* replicate each other, and the techniques of faceted classification can be and have been applied to thesauri. Although the NISO standard does not discuss this, a thesaurus may indeed adopt notational codes to identify concepts and then associate all that concept's labels—both preferred and unpreferred—with the concept identifier. Using a notation for concept identification might not be common for thesauri, but neither is it unusual, particularly for complex systems with many levels of hierarchical relationships. Indeed, Soergel discusses such possibilities in his 1974 manual. If most thesauri do not employ notational codes to represent concepts, the reason is that creating and managing a separate system of concept identifiers is a resource-intensive task, particularly when labels and their relationships must also be maintained (a *classification*, which does not concern itself with labels, does not have the latter problem). And so, although the creation of a separate system of identifiers to track concepts is not unknown, now or in the 1970s, it is more common to rely on the *preferred term* for this function. Importantly, using a notation is considered a formatting decision rather than a conceptual one. For someone like Soergel, there is no fundamental difference between expressing a concept via a coded identifier and expressing a concept as a preferred term. In other words, the distinction between concepts and labels is clear in the minds of thesaurus designers, even if the language of terms makes it seem otherwise.

To further support this theory, we can look at the development, just subsequent to the release of the 2005 NISO standard, of the Simple Knowledge Organization System (SKOS). SKOS is a schema for defining, labeling, and relating concepts by means of Resource Description Framework (RDF) statements. RDF is a way of encoding statements about resources, such as "Concept X is labeled with the term 'Variety'" or "Concept X is broader in meaning than Concept Y." RDF is associated with the World Wide Web Consortium (W3C) Semantic Web initiative, also known as Linked Data.

If that all sounds like alphabet soup, just think of SKOS as a way to express thesauri and other controlled vocabularies in a format that computers can readily process and that can be easily aggregated with other datasets. SKOS uses the concept as its basis; labels are associated with concepts. The language of terms is avoided.[12]

SKOS was developed through extensive collaboration between computer scientists and indexing experts (Baker et al. 2013). However, SKOS was designed to facilitate the expression of "lightweight ontologies" for a variety of purposes—not just controlled vocabularies for indexing and retrieval. Accordingly, the language of terms was felt to be too vague and imprecise for SKOS, and it was omitted. In turn, some welcomed SKOS as an opportunity to clarify the language of controlled vocabularies more broadly. For instance, the 2011 ISO standard for controlled vocabularies, ISO 25964, emphasizes the primacy of concepts, characterizing terms solely as labels. Unlike the 2005 NISO standard, ISO 25964 states forthrightly that "concepts are represented by terms, and for each concept, one of the possible representations is selected as the preferred term." ISO 25964 also includes a data model diagram for controlled vocabularies, strongly influenced by SKOS, that puts the concept at the center: terms in this data model are explicitly and clearly nothing more than labels for concepts.

From this account, the simple answer to my question about terms seems hard to refute. It seems abundantly clear that, with regard to long-established traditions of data management, labels are clearly separate from, and subsidiary to, concepts. In this orientation, the catalogers, classificationists, and indexers of times past share a similar perspective with the data scientists and ontology engineers of the present. If we can rely on this consensus, then we can feel secure in changing the names of concepts whenever and for whatever reason that we like. If concepts are independent of their labels, then choosing a different label doesn't affect the meaning of the concept at all. *Variety?* Poof, become *Cultivar* instead! The process is simple and mechanical. A bad name can be fixed by purely technical means: search and replace.

Empirical Realities of Extracting Concepts from Documents

To be sure, it does seem reasonable that an apple variety remains the same no matter if what we call it: Cox's Orange Pippin or 78B. Nonetheless, the simple answer to my question about terms—that an apparent fuzziness between terms and concepts arose because indexers in the 1960s didn't have the technical infrastructure necessary to manage concepts and labels separately—might be a little *too* simple. Let's think for a minute about this idea of *concepts*. How do we come to understand concepts like the notion of a Cox's Orange Pippin? Some of our understanding arises from direct experience: from eating, growing, or cooking those apples. But another aspect of our understanding is rooted in discourse: in reading and talking about apples, using certain names to do so. Is it really so easy to separate this understanding from the names we employ? What's more, our empirical understanding and our discursive understanding contribute to and contaminate each other. We can't easily disinvest our knowledge from the words we use to express it.[13]

In the adventure essay for this chapter, I assert that *Yellow*, *Oriental*, and *Asian* are wholly different concepts, rather than synonymous labels. That's a direct and dramatic example. But even apple varieties may seem different with different names. The Mutsu apple was bred in Japan in the 1930s, but when it was exported to the United Kingdom in the 1960s it became known there as Crispin. *Mutsu* in Tokyo, *Crispin* in London: the same apple, two names, two locations. The same identity? In some ways, yes; of course, yes. It's the same apple cultivar. But in other ways, no; actually not. The two names mark out different stories, different ways that people in different places have experienced the apple and understand it in relation to similar concepts. And so, the complicated answer to my question about terms is that the ambivalence and uncertainty of the language in the 2005 NISO standard might be more precise and accurate—and less accidental—than otherwise.

It's telling here to look at the practical work of vocabulary creation. Practice manuals and standards documents fill many pages with meticulous explication of small technical areas, such as term syntax and formatting. But guidance is skimpy regarding the identification, definition, and naming of concepts (or preferred terms) in the first place. The 2005 NISO standard says almost nothing about this. Soergel, for his part, describes the collection of terms as a primarily technical process, in which all but the most generic and abstract of noun phrases are meticulously copied from source documents and recorded on a specially designed 5-by-8 inch cards. Concepts are identified by sorting and merging similar cards, and are then expanded by grouping related cards and identifying the groups. In this depiction, the identification of concepts via terms in source documents, although performed by humans, seems almost computational. One can glimpse here how the language of terms may have seemed perfectly acceptable to someone as rational as Soergel. In his characterization, source documents are essentially repositories of encoded concepts—quantitative data more than linguistic expression. Although the language of source documents might vary, this variation is a matter of symbolic representation rather than sense. The difference between *Mutsu* and *Crispin* is the same as the difference between *Cox's Orange Pippin* and *Coxs Orange Pippin*.

Soergel's presentation of the term collection/concept identification process as fundamentally logical, rather than hermeneutic, is not unusual. Library classificationists of the early twentieth century approached their work this way also, in the mode of scientific discovery rather than creative design.[14] So too did the creators of indexing languages in the mid-twentieth century, much like the data modelers of today.[15] Across these domains, concept identification has most often been characterized as a process of extraction from source data, as achieved though systematically applied analytic techniques (whether performed manually by people or automatically by computers).

Implicit within this orientation is the key role of source material selection. Different sets of primary data result in different terms being harvested. This, then, becomes the focus of design. The various design rationales put forth in the literature of classification and indexing, commonly expressed as types of *warrant*, are fundamentally criteria for source selection (Beghtol 1986). For instance, *literary warrant* (what has been written), proposes that source documents comprise the collection for which an organizing system will be used (when literary warrant was proposed by Hulme (1911), this was a library, but it could be any collection—the Web, for instance). *User warrant*, on the other hand, focuses on what information seekers look for, and so the source material might include search queries as well as the most frequently accessed documents. *Scientific warrant* centers on what subject-matter experts currently believe, and so source documents are those that contribute to current scientific consensus. Once the source material has been assembled, however, subsequent processes of term collection and concept identification proceed similarly. Whatever the source data might be, a rational sorting process is applied, and concepts issue forth. (Conceptually, this process is very similar to the process for training machine learning algorithms on representative sets of test data.)

In my own experience, there is certainly a level at which the characterization put forth in the literature—of concept extraction as a kind of logic puzzle, in which source documents are manipulated until a coherent solution shakes out—does obtain. But there is an additional layer that precedes the puzzle part: the human reader, reading texts written by other humans. In the procedures described by Soergel and others, source materials are data to be mined. *Reading* the source materials is portrayed as unnecessary or even wrong; criteria for collecting terms should come from somewhere else (e.g., from scientific consensus for scientific warrant or search queries for user warrant) and not from my own interpretation of the documents. But, well . . . that's hogwash, isn't it? And it's not just that I'm a human and I can't resist reading the materials I'm supposed to be mining. It's that reading them is *necessary*. If I don't read them, I miss a good deal of their data—the data that arises from the interplay of terms rather than the application of concepts.

Are *Crispin* and *Mutsu* the same? If I am a concept extractor playing a logic puzzle, yes. If I am a reader interpreting accounts arising from divergent modes of discourse, then *maybe* not. As for what to do, the confusing, ambivalent, and superficially unsatisfying language of the 2005 NISO standard fits this reality surprisingly well. The world of concepts and the world of labels are not separate; they swirl around each other and, as terms, take on different characters depending on what I do with them. As the creator of a controlled vocabulary, whichever decision I make will be partially accurate: whether I take the terms as equivalent and arbitrary or whether I take the terms as unique and meaningful, whether I find their referent to be the same concept (*Crispin* equals *Mutsu*), or a different concept (*Crispin* does not equal *Mutsu*), or both the same concept and a different concept at the same time (*Crispin* does and does not equal *Mutsu*).

I realize that what I'm saying here sounds vague. Of course it is. Terms are impossible to pin down and also easy to pin down; meaningless and important at once. We can't get out of this duality merely by asserting that names matter, or that they don't. Deciding when and how names matter—and when they don't matter—is both a necessary task and an impossible one. If we make it seem any less complicated and contingent—any less human, in other words—we're only fooling ourselves. Names are not problems to be solved, with solutions technical or otherwise. Rather, names represent situations to be contingently negotiated.

Reconfiguring Concept-Label Relationships

A few years before Dagobert Soergel published his thesaurus construction manual, library cataloger Sanford Berman released *Prejudices and Antipathies, a Tract on the LC Subject Heads Concerning People* (1971). Berman's book comprised a set of Library of Congress subject headings that presented certain groups of people as lesser or inferior. (In library cataloging, *subject headings* are the preferred terms used to describe a work's aboutness. The Library of Congress makes its catalog data available to other libraries, and almost all libraries in the United States use LC data as the basis for their own catalogs. Therefore almost all libraries in the United States use LC's controlled vocabularies, even when they create some of their own data.) For each objectionable subject heading—Yellow Peril is one example—Berman explained how the heading unfairly characterized the group in question and provided suggestions for the heading's replacement.[16] Berman's text could not be more different from Soergel's: although it is likewise carefully compiled, it is vehement and passionate. Berman was an idealist, clearly motivated by issues of social justice, whereas Soergel was a rational problem solver for whom political motivations were irrelevant. However, despite these very evident differences, Berman and Soergel appear to share certain fundamental premises about data: how we create and manipulate it.

Berman's introduction presents an episode with the information scientist A. C. Foskett, a known expert in subject indexing. (In this vignette, Foskett serves as a proxy for Soergel; they both represent a logical/computational perspective on knowledge representation.) Foskett was asked to comment on the "racist/colonialist bias" of the Library of Congress subject headings (LCSH). Foskett replied that LCSH reflected the "historical bias" of its source material: the holdings of the Library of Congress. Precisely so, pounced Berman: "Once recognized, surely the most foolish and wrong-headed aspects of the bias can be corrected." Berman continued,

> Just because, in short, "we were brought up that way" is no valid reason for perpetuating, in our crania or in catalogues, the humanity-degrading, intellect-constricting rubbish that that litters the LC list. Moreover, within the context of a world increasingly polarized by White/Black, rich/poor, West/Tiers-Monde, the burden is fully and immediately upon us to at least rectify the worst features of library practice. (Berman 1971, 16)

If you agree with Berman's position (as I do), you will find his takedown awesome (as it is). But Berman's diagnosis of a problem (the subject headings are biased) and subsequent rationale for a solution (eliminate the bias) was, perhaps, not radical enough. The way that I read this exchange, Berman and Foskett remained in alignment about data design fundamentals. Their disagreement can be explained as a quibble: a dispute over selection criteria for source data. Foskett's response demonstrated his support for literary warrant: aboutness data should emerge from the collections being described. LCSH represent the holdings of the Library of Congress. If those documents include certain concepts under particular labels, then that is just what the data says. Berman's rejoinder countered that this source data was flawed ("biased"). Documents from the past do not represent our understanding of the world today. Berman proposed another kind of source data, based in the current world and the way that we see it (as opposed to the way that "we were brought up"). One could see Berman's proposal as a variation of scientific warrant, in that it would eliminate source documents that are "foolish and wrong-headed"; or one could see this as a kind of ethical warrant, in which documents that subscribe to certain ethical codes (that are not humanity-degrading) would be included, and documents that do not subscribe to those ethical codes omitted. (I have the impression that Berman equated these warrants—that he believed that the scientifically valid ones [not foolish] were also those that promoted certain ethical principles [that are not humanity-degrading].)

But Berman did not suggest alternatives to existing processes of concept identification and labeling—as laid out, for instance, in Soergel's manual. He did not offer a different mode of understanding concepts, labels, and their relations—of dwelling differently within the world of terms. In my reading of Berman's manifesto, a vocabulary creator would perform similar actions to those proscribed by Soergel—and make similar decisions using similar criteria. Of course, the outcomes of the vocabulary creation process would be different, because the source data would be different. Racist concepts such as Yellow Peril would no longer appear in LCSH, under that label or any other label.

However, if we don't allow for different kinds of relationships between concepts and labels—for different ways of accounting for name data—then continually changing our source material to align with current principles (for scientific validity, for ethical implications, for social responsibility, and so on) might merely displace some problems for others. To explain what I mean, let's look at one of the problematic headings that Berman identified in *Prejudices and Antipathies:* Yellow Peril. What would happen when LCSH, following its new scientific/ethical warrant, abjures the existence of Yellow Peril? How does the aboutness data change for books formerly described with that heading, such as Sidney Gulick's *The American Japanese Problem: A Study on the Racial Relations of the East and West?* In *Prejudices and Antipathies*, Berman recommended the heading East and West as a substitute for Yellow Peril. On the surface, this substitution seems reasonable for Gulick's book. *East and West* is right there in the title, after all, and *Yellow Peril* is one way of describing relations between East and West.

But changing the subject heading wouldn't change Gulick's book. As the adventure essay for this chapter discusses, Gulick did endeavor to promote understanding and sympathy between American and Japanese societies. At the same time, however, his book accommodated American racism and used, or even extended, its terminology. When Gulick's book was described with the term *Yellow Peril*, it was associated with other works that similarly accommodated or endorsed such views. But when Gulick's book is described with the term *East and West*, it becomes newly associated with books that discuss East and West in a variety of ways, from a variety of perspectives, using different forms of discourse. For instance, Edward Said's *Orientalism* (1978) is described with the heading East and West. Describing these two books with the same aboutness data is not inaccurate. However, the impression of equivalence that now obtains between these works is uncomfortable. Said, using the term *Orientalism*, critiqued the very ideas that Gulick, using the term *Yellow Peril*, pandered to. Different terms: *Orientalism* and *Yellow Peril*. Same concept: East and West. Under Berman's regime, aboutness data hides the opposition between Said's book and Gulick's book, rather than revealing it.

Is there a solution to this problem? Characterizing it as a problem implies an answer. I don't think that we can *solve* name data; we can only decide how to account for it in different situations. In the Yellow Peril situation, if I had the magic wand that I talked about in this chapter's adventure essay, I would try to maintain the conceptual link to East and West while likewise including the terminological link to Yellow Peril. For Gulick's book, I might retain but re-situate the term for Yellow Peril, keeping it but explaining it: for instance, in the form *Yellow Peril (a racist idea about Asians)*. But I would likewise retain the term *East and West*, which would also preserve the connection to Said's book. In taking these paired actions, I would hope to create aboutness data that shows, at the same time, how the name *Yellow Peril* matters a lot *and* how the name *Yellow Peril* doesn't matter at all.

If we take a traditional perspective on data management, my non-solution proposal about Yellow Peril is awful. It is biased, because it arises from my socially situated interpretation of the source data. Moreover, it can't be reduced to easily operational rules that can be applied across many equivalent situations. My proposal might make sense for Gulick's book and for the term *Yellow Peril*, but it can't be extrapolated into some universal rule.

Well, both of these objections are perfectly valid, if our goal in creating data is to enable efficient processing by computers. But is that what we want? Bias, for instance, is often portrayed as terrible, but when humans talk about bias, what they are often talking about—what Berman is talking about, for instance—is not bias but unfairness. We can remove bias from data only when we remove all traces of humanity; but we can achieve fairness only when we insert more humanity into our data. What this means, unfortunately, is that name data cannot be *solved* computationally, but only debated—politically.

7
LOCALITY

This chapter contemplates position: how our location within a culture affects data design and implementation. Being located within a particular environment is conceptualized as integral to data quality rather than antithetical to it.[1]

In the adventure essay, I tell stories about

- A poet talking about American racism.
- Height measurement at the doctor's office.
- Race and ethnicity in the US Census.
- Race and ethnicity in online dating sites.
- A novelist talking about Danish racism.

In the reflection essay, I ruminate over

- Constraint and creativity in cognitive categorization.
- Local conditions and cognitive prototypes.
- Historical efforts to standardize data collection across locations.
- Data creation as an assertion of values (an extended example from library cataloging data).
- Venerating the humanity in our data.

ADVENTURE: HAVE YOU EVER BEEN TO LOUISIANA?

A Poet Talking about American Racism

In one of my Danish language classes, our textbook included a dialogue between two recurring characters, Jonas and Gertrud. The dialogue was called "Have you ever been to Louisiana?" Would Jonas and Gertrud be taking a holiday in the United States? I wondered. But of course the dialogue referred to the Louisiana Museum of Modern Art in Humlebæk, about 45 minutes north of Copenhagen. It had nothing to do with beignets or bayous.

When we studied that dialogue in class, I was fully aware of the existence of Louisiana (the museum). I had even become a museum member so that I could visit anytime I wanted. I was far more familiar with Louisiana, the Danish museum, than I was with Louisiana, the US state. Nonetheless, every time I looked at the title "Have you ever been to Louisiana?" I was momentarily disoriented. I felt, I guess, very foreign, even in a class composed entirely of foreigners. My classmates were all from Europe or Asia, and no one else was thinking about New Orleans when they looked at the title of that dialogue.

Weirdly enough, I had felt unsettled in exactly this way during my previous visit to Louisiana (the museum). In August 2019, I attended Louisiana Literature, a yearly festival with authors from around the world. At the festival, I went to a session with the American poet Claudia Rankine, interviewed by the Danish journalist Synne Rifbjerg. As Rankine summarized themes in her current work, I nodded in recognition. Rankine's account of a doggedly racist, politically stagnant United States was grimly unsurprising. Nonetheless, something about that conversation, at Louisiana, felt strangely askew, as if reality had been transposed onto an alternate dimension. This feeling of oddness didn't seem to have anything to do with the actual words being spoken, and yet it seemed very much wrapped up in the words being spoken. What was it?

As I mulled this contradiction, I found myself becoming increasingly aware of my surroundings in the Louisiana concert hall. The audience was certainly homogeneous, with its primarily White, well-tended faces. But this kind of homogeneity would not be unusual in the United States either, despite our pride in diversity. When I lived in Austin, Texas, there was a yearly book festival at the state capitol. The Austin auditoriums were darker and stodgier than at Louisiana, but the audience composition there, too, leaned White and bourgeois—more than one might anticipate, given the city's population.[2]

And yet, I began to realize, these feelings of strangeness *did* arise from a sense of displacement. Neither the audience nor the subject matter were alien to me, but their conjunction in this place felt different. The matters being discussed on stage—rapid gentrification, persistent gun violence, incarceration of the Black population—were part of everyday life in my American city of Durham, North Carolina. But to the people around me, Claudia Rankine's account was like a newscast from a distant war zone. To be sure, we all processed the basic content of that on-stage conversation similarly, namely, that racial inequality in the United States continues to increase. This was clear from the responses of the interviewer and later from audience questions. But the Danish viewers were positing an elsewhere. I wasn't.

I began wondering what it would be like if I were listening to Claudia Rankine speaking the same words in a different place, maybe somewhere at home, in Durham. But where in Durham? At first I imagined her at the John Hope Franklin Humanities Institute on the Duke University campus. Then I mentally transported her a few miles away, to North Carolina Central University. Duke is a wealthy, private university. NC Central is a public HBCU (historically Black college or university). It was easy to imagine both audiences as primarily Durham residents, who might agree, Yes, I recognize what Claudia Rankine is talking about, here in our community. But in the Duke audience, more people would come to that understanding from this perspective: I am concerned about inequality in my neighborhood, even as I am also a White gentrifier whose presence is squeezing non-White residents out of it. (I would be one of those people.) In contrast, in the NC Central audience, more people would come to that understanding from this perspective: I am concerned about inequality in my neighborhood because the presence of White gentrifiers is making it uncomfortable for me and my family to continue to live here, socially and economically. Same words, same speaker; different settings, different audiences. Same data?

Height Measurement at the Doctor's Office

On the one hand, this seems like an absurd question. Data is supposed to be factual, and facts aren't supposed to change depending on where, how, and for whom they are generated. I am a certain height: the environment in which I am measured doesn't have anything to do with how tall I am. And yet, the last time I went for my yearly physical examination, the medical assistant measured my height as usual and said, "You are 5 feet 2 and 1/2 inches." "What?" I said. "No, I'm 5 feet, 3 inches. Right?" For more than 30 years, whenever I had been measured, I had been told that I was 5 feet 3 inches. I stood up a little straighter. "Are you sure?" I said. The assistant verified her measurement.

Did I shrink? I don't think so. There is an obvious and mundane explanation. Most of the people collecting my height data over the years were rounding the number upward to the nearest inch. But why round to the inch? The height rods on physician scales are commonly incremented in eighths of an inch, not whole inches or half inches. Again, the explanation seems banal. It was standard practice in the environment in which I was measured—that is, in the United States, in a general practitioner's office in 2019—to express height in integers of feet and inches. Moreover, it was also conventional, in the environment in which I was measured, to be generous in rounding up. Many people prefer the idea of being taller. If I were a medical assistant measuring people, I would round up! Probably many of us are not quite as tall as we think that we are. (If you are used to being measured in centimeters or some other way, your sense of height will be different, of course. Keep your own experience in mind for the next few paragraphs.)

The way that I tell this story may make it seem that height data in doctor's offices is controlled by the people doing the measuring, and that the social explanation—generous rounding—is therefore the determinant one. If this were true, it might be reasonable to infer that technical decisions, like choosing a particular system of measurement units, operate independently of social ones, like choosing to round up in borderline cases. The action of the most recent medical assistant, to assert that I was actually shorter than I thought, would seem in line with this interpretation. The agency clearly lies with the assistant, whose judgment may be clogged with subjectivity and dependent on environmental conditions—the assistant, after all, is telling the patient the result rather than recording it silently in a database.

Such an interpretation, however, understates the role of the selected units in making the social decision possible. Measuring in feet and inches, in whole integers, is part of what enables rounding up. This particular system of units both encourages us to want that extra inch and makes it easy for us to get it. When we measure in feet and inches, each inch seems to matter a lot. We want to round up! At the same time, being within the range of a rounding decision is so minor, it's easy to be generous with it. It's almost like the choice to measure in whole inches sanctions us to add a little. Of course, it could easily work the other way around: it is because that inch seemed to matter that we settled on measuring the way that we do. It's impossible to separate a desire for that extra inch from the way in which we express the total.

Indeed, it's fun to contemplate how minor technical changes might affect interpretation of height measurements. For instance, we could express the measurement solely in terms of one unit (feet *or* inches) rather than two units (feet *and* inches). So instead of being 5 feet 2 inches tall, I would be 62 inches tall, or 5.21 feet. As someone used to being measured in feet *and* inches, 5 feet

2 inches seems terribly short. But 62 inches or 5.21 feet . . . meh, I don't care so much. And if we changed the units to centimeters? To me, an American used to feet and inches, the difference between 160 centimeters or 157 centimeters is essentially meaningless.

An automated process of height measurement seems like an easy solution here. Just get rid of the soft-hearted (or hard-hearted) medical assistants. But the agency of the human assistant was always indebted to the technical apparatus, of which the decision to use particular units is a key component. In other words, any soft-heartedness is not just in the human hand and eye recording the value: it's also in the choice of measurement scale. If we want to emphasize precision, changing the measurement scale to use smaller units (millimeters, for instance) will have a greater effect than automating a feet-and-inches arrangement. Both the sympathetic human hand and the emotionless machine hand will measure more precisely in millimeters and, more importantly, a continued practice of measuring in millimeters may well leach the sympathy from the human hand, should it become clear that the measured subject won't cling as dearly to the odd millimeter.

So, why do Americans tolerate the vagueness of feet and inches when we could so easily have millimeters? We seem to prefer it. When I snarkily observed that most of us are probably not as not as tall as we think we are, that was misleading. We are exactly as tall as we think we are, given the somewhat elastic form of data that we have communally chosen. If anyone was wrong about how tall I am, it was the conscientious assistant. In an American doctor's office, of course I am 5 feet 3 inches, and not 5 feet 2 1/2 inches, 62 inches, 157 centimeters, or anything else. The environment in which I am measured *does* matter. Data collection is not a universal enterprise. It's a sociotechnical system of exquisitely related components, and we can't separate the what and the how from the who, where, and when.

When I was in the auditorium at Louisiana, thinking about permutations of *them* and *us, here* and *elsewhere*, I found myself returning again and again to the mundane technical decisions that might enable and reinforce those impressions. I don't actually grow or shrink when I cross borders, of course, but am I more likely to be measured as 5 feet 3 inches in the United States and 157 centimeters in Denmark? Then *I* don't shrink in Denmark, but my *height*—if we take height to be the data of how tall I am—changes nonetheless. Height has a local dimension.

If we found the local aspect of height to be distressing—perhaps because we wanted to aggregate height data and compare it globally—we could easily imagine this as a straightforward problem to solve. It's not hard to envision a standardized regimen to regulate local variations, at least for some specialized purposes. As much as everyday measurement in doctor's offices might vary, we could still establish, say, scientific measurement for research purposes via a universal height standard that includes specific units, data collection and rounding protocols, and so on. The distance between *here* and *elsewhere*, in terms of height data—if we can't eliminate that gap, we should be able to narrow it. (Chapter 4 provides another perspective on such efforts, in the context of reproducibility and interoperability.)

The conversation at Louisiana, however, revealed a different kind of gap. The Danish audience didn't merely perceive *racism in the United States* differently from the way that I did. The Danish audience perceived *race itself* differently from the way that I did. Race, even more so than height, is a challenging concept to define in the abstract. The way that we understand race tends to be bound up in how we implement it as data. And that implementation is intensely local.

Race and Ethnicity in the US Census

In the United States, race data is a standard component of demographic data, like age and sex. Race data is collected for population statistics, for workplace statistics, for health statistics; race data is ubiquitous.[3] And yet it is not defined in a standard manner (unlike, for example, measuring height in feet and inches). Although race is typically operationalized as selection from a controlled vocabulary, data values in those vocabularies vary across sites of collection.

Because government agencies often report demographic statistics as a part of their public mission, one might expect a government standard to be a model exemplar in terms of how race data is collected. In particular, the US Census, conducted every ten years, is supposed to compile an accurate, comprehensive picture of US residents and their distribution over the country. Census data is relied upon to support policy decisions, program assessments, and academic research, among other uses. However, as chapter 4 observes, many people find the census race questions to be profoundly confusing.[4] It is nonetheless instructive to consider how the census implements race data, especially if we compare the current version (used in 2000, 2010, and 2020) to a proposed future version, which was recommended for the 2020 census but ultimately not implemented (US Census Bureau 2017). Table 7.1 shows

Table 7.1
Potential data values for race in the US Census

Version A Used in 2000, 2010, 2020[5]	Version B Proposed for 2020 but not implemented
Race (Can select multiple values) • White • Black/Afro Am, or Negro • American Indian or Alaska Native • Asian Indian • Chinese • Filipino • Other Asian • Japanese • Korean • Vietnamese • Native Hawaiian • Guamanian or Chamorro • Samoan • Other Pacific Islander • Some other race Spanish/Hispanic/Latino origin • Mexican, Mexican Am, or Chicano • Puerto Rican • Cuban • Another Hispanic, Latino, or Spanish origin • Not Hispanic, Latino, or Spanish origin	Categories (Can select multiple values) • White • Hispanic, Latino, or Spanish origin • Black or African Am • Asian • American Indian or Alaska Native • Middle Eastern or North African • Native Hawaiian or other Pacific Islander • Some other race, ethnicity, or origin *(More specific categories can be written in under one of these choices. For example, someone who selects White could additionally write Finnish; someone who selects Asian could additionally write in Thai.)*

these two different versions of census race categories, identified as version A (2000–2020) and version B (Proposed).

A comparison of versions A and B can help us to understand the local character of race, as implemented in these versions of the census.[6] Here's how political scientists Jennifer Hochschild, Vesla Weaver, and Traci Burch characterize version A:

> The first question implicitly asks about ethnicity—but only for Hispanics and, curiously, Spaniards ... the second question first defines race as a color—"White" or "Black." But there is no Brown, Yellow, or Red; instead the answer category shifts to race as a tribe—but only for Native Americans. Race then appears as a nationality—but only for nations that are, roughly speaking, in South or Pacific Rim Asia. Finally, the question gives up. (Hochschild, Weaver, and Burch 2012, 4)

What Hochschild, Weaver, and Burch observe is that these data values mix different sorts of characteristics. Race, in this data implementation, might variously pertain to color, tribe, nationality, or *other*. (In the language of chapter 5, these values employ different principles of division.) Meanwhile, Hispanic ethnicity is a separate construct entirely, distinct from race.

Version B changes significantly. It combines all the data values into a single set, including those for Latinx origin. (To accommodate this change, the words *race* and *origin* are removed as broader terms, and the set of data values is labeled as *categories*.) Additionally, different Asian origins, which had been listed separately in version A, are collapsed into a single value. A completely new choice, without precedent in version A, is added: Middle Eastern and North African. As a final change in version B, there is an option to provide a more specific value, of any nature, to any selected main category. So, for instance, someone who selected *Hispanic, Latino, or Spanish origin* might write in *Cuban*.

However, although version B differs significantly from version A, its data values still involve a mix of characteristics—the same characteristics that Hochschild, Weaver, and Burch identify for version A. In version B, race is sometimes a matter of color (for White or Black), sometimes a matter of sociocultural affiliation or heritage (for Hispanic, Latino, or Spanish origin and American Indian or Alaska Native), and sometimes a matter of nationality or geographic origin (for Asian, Middle Eastern or North African, or Native Hawaiian or Pacific Islander). Version B, then, maintains version A's ambiguity about race with quite profound consistency. The coherence of version B's implementation is perhaps best encapsulated by the new label for its *other* data value: *Some other race, ethnicity, or origin*. On the one hand, this label obscures the construct being datafied as an amorphous combination of race, ethnicity, and origin. On the other hand, it cleanly delineates and separates those constituent concepts. One of the implications here is that the values for White and Black, with their basis in color, are foundational to race, as implemented in the census. The relative stability of the White and Black values across versions lends credence to this. White, in particular, is central; uniquely, nothing about it changes from version to version.

Race and Ethnicity in Online Dating Sites

The tension that version B highlights between race, ethnicity, and origin—a data implementation that consistently attempts both to distinguish and elide these notions—is common in other US settings in which race data is collected. When I teach students about data design, I often use examples from online dating services. It's instructive to compare these variations also in relation to the census.

Table 7.2 displays available data values to define one's race in six dating services.

With the exception of JDate, these implementations of race data show an assemblage of characteristics similar to that exhibited by the census. Once again, race is variously defined by color, by sociocultural affiliation or heritage, or by geographic origin. Similarly to the census, these dating sites consistently invoke the same set of characteristics as constituent of race, differing primarily in minor details. In particular, the White/Black core is quite stable, and the White data value is the most stable of all. (The labeling of the White data value as Caucasian is deceptive; in the United States, Caucasian is a synonym for White, and it has nothing to do with the Caucasus region.)

JDate, which does not collect race or ethnicity data, provides a distinct complement. JDate describes itself as "Jewish singles looking to make a connection with other Jewish singles." Jewishness, presumably, takes the place of other race and ethnicity data. But is additional data unnecessary because Jewishness is considered to be its own race or ethnicity, or is additional data unnecessary because all Jews are presumed to be members of some other, more general race or ethnicity, such as White or Mediterranean? Other dating sites do not include Jewish as a data value, which may imply the second interpretation. However, other dating sites do include Middle Eastern and Latino as potential data values. Sephardi and Israeli Jews might well select those values, rather than White.

So it's ultimately uncertain whether Jewishness in JDate is a primary category (like Latinx origin in version B of the census) or a secondary category (like Latinx origin in version A of the census), or both, or neither (that is, a religion only). Indeed, what is really noteworthy is not the ambiguity of Jewishness within the context of JDate, but the lack of ambiguity regarding Jewishness everywhere else. Jewishness has, of course, been characterized along racial lines in many societies.[7] To me, this treatment of Jewishness as incidental to race underscores the prominence of Whiteness in American race implementations. In the United States, Jews are White primarily and Jewish incidentally, at least as the implementation of race data is concerned.

Table 7.2

Potential data values for race in six online dating services (as of July 2020). Of the six, Match and OK Cupid serve a general audience. Black Planet and JDate are targeted toward Black and Jewish daters, respectively. Dharma Match focuses on people interested in spirituality, and Gluten-Free Singles targets particular dietary requirements. Four of the dating services label this construct Ethnicity rather than Race, but the available data values are similar.

Match	OK Cupid	Black Planet	JDate	Dharma Match	Gluten-Free Singles
Ethnicity (Can select multiple values) • Asian • Black/African descent • East Indian • Latino/Hispanic • Middle Eastern • Native American • Pacific Islander • White/Caucasian • Other	Ethnicity (Can select multiple values) • Asian • Middle Eastern • Black • Native American • Hispanic/Latino • Pacific Islander • Indian • White • Other	Race (Can select multiple values) • Asian/Pacific Islander • Black/African American • Hispanic/Latino • White • Other Ethnicity • Brazilian • British • Cuban • Dominican • Eritrean • Ethiopian • Ghanian • Haitian • Ivoirian • Jamaican • Nigerian • Panamanian • Puerto Rican • Senegalese • Trinidadian • Other • South African	*No data collected for race, ethnicity, or any similar construct.*	Ethnicity (Can select multiple values) • African/Black • Asian • Caucasian • East Indian • Hispanic/Latino • Middle Eastern • Native American • Pacific Islander	Ethnicity (Can select only one value) • African • African/American • Asian • White/Caucasian • East Indian • Hispanic • Indian • Latino • Mediterranean • Middle Eastern • Mixed

In scrutinizing these kinds of data implementations, one can be tempted to focus on their perceived deficiencies. We might, for instance, complain that version A of the census is poorly designed because it doesn't make sense to describe Latinx origin as distinct from race. Or we might observe that Black Planet includes Nigerian and Ethiopian as ethnicities but not Kenyan or Somali, and tut-tut over what we see as arbitrary choices. Often, such complaints focus on individual data values—whatever seems wrong. But a focus on individual incongruities can miss how the system functions holistically. Each set of data values that we've looked at from the census and from the online dating services comprises a tightly integrated and compressed block of expression—like a poem. We would never look at a single line of a poem and think, bah, I don't immediately understand what that line means or why it is there, so clearly it needs to be rewritten. We recognize that we need to read each line of a poem in relation to the whole.

Moreover, we read a poem in relation to our understanding of poetry more broadly: in the context of other poems in a literary tradition, in the context of literary techniques and genres, and in the context of a particular author's body of work. A set of data values is no different. Accordingly, it is useful to direct our attention, at least initially, toward the functional logic of a data implementation: what makes it coherent, as opposed to what seems incoherent.[8] In all these implementations of race data, in online dating and the census, the same set of characteristics appear both ambiguously *and* consistently. For such an uncertain and dynamic concept, race is implemented in strikingly uniform ways.

We can extend such an explanation by incorporating a practice-oriented perspective—how data is created with the available data values. In my teaching, for instance, I might ask students to take on the perspective of an online dater, creating a profile and using the profiles of others. Which data values would they select and why? Would it be a good idea, I ask, to create a standard set of data values for all online dating services, so that we could aggregate all the profiles into a single dataset, a kind of union catalog of daters?

I generally get two types of answers to these questions. The first responses tend to focus on perceived inaccuracies. They might observe, for instance, that White seems like a race, but Middle Eastern seems like an ethnicity. Are these data values supposed to be races or ethnicities? students will ask. These students believe that it is important to define these concepts correctly and precisely, so that people searching the dating profiles will not be surprised by the results. They will make clarification suggestions like, White is a race, Middle Eastern is an ethnicity, and Saudi is a nationality. When concepts like race, ethnicity, and nationality are ironed out, it will be possible, these students hope, to standardize the data and aggregate it. Which would be a great idea, because, of course, isn't it better to have as many matches as possible?

But then a second group will begin objecting to this plan. Creating a profile for an online dating site, these students contend, is not a matter of referring to precise definitions of race and ethnicity. To create a dating profile, you think about what you want to convey to other members of the dating site, given the available choices. A dating profile is a form of flirting. It's rhetorical. A Pakistani Muslim student, for instance, might select Middle Eastern in addition to, or instead of, East Indian, depending on the data values that existing users for a particular site have already selected. The student's choice here is not based on whether Pakistan is "really" in the Middle East, or "really" part of the Indian subcontinent, or "really" both; nor is it based on whether "Middle Eastern" or "East Indian" are supposed to identify locations or physical similarities or sociocultural similarities—whether these terms reference *race* or *ethnicity* or what those constructs might really be. None of that matters; all that matters is what kind of signal a choice of data value is perceived to make for a certain user community. These students are thinking a little more holistically: how each dating site works as a local data environment. And so, for these students, a union catalog of daters is not necessarily a wonderful idea. A person selects a particular site like Black Planet or Gluten-Free Singles purposefully, and constructs a data profile purposefully, with a specific audience in mind. Aggregating the data does away with all that careful situatedness.[9]

At this point, the first set of students looks a little bit uncomfortable. Choosing a different data value for race depending on different social circumstances seems wrong, but the second set of students clearly has a point: dating sites are about the alignment of personas in a complicated dance. So I give the students another context to think about: the census. I show them versions A and B, and I ask about the progression from one version to another. What about selecting a data value for the census, as opposed to an online dating profile? Is it more important to think about accuracy and precision?

Now the second set of students are wrinkling their foreheads. All of a sudden the stakes seem different, and they are doubting their initial stance.

The census data clearly needs to be correct and reliable; it needs to reflect the actual composition of the population. Otherwise, policy decisions will be based on faulty evidence. This leads to direct material consequences: political representation can be skewed, public resources allocated differently, and our understanding of social disparities altered. Often the students hesitate, not knowing how to respond.

To get the discussion moving again, I might offer a hypothetical situation. Let's imagine two different people, Y and Z. Y and Z are both biracial, with one Black parent and one White parent. Both Y and Z are physically ambiguous; they could be taken for White or Black, appearance-wise. Y and Z are filling out the 2020 census. Y selects Black. Y makes this selection because Y experiences life as a Black person, given Y's skin color and features. Further, Y feels that it is important to ensure that BIPOC are not undercounted in the census, given that non-White communities are generally less likely to receive and return census forms. Z, on the other hand, selects both Black and White. According to Z's rationale, having one Black parent and one White parent dictates this choice. I ask, Has either Y or Z entered incorrect data?

The class chews its collective fingernails. What are the instructions on the census? someone might ask, stalling for time. The instructions are vague, I explain. In 2010 and 2020, respondents were directed to select a race, with further clarification to mark one or more boxes. Elsewhere, the Census Bureau explains that its race and ethnicity data is based on self-identification. If self-identification is the prescribed mode of data implementation, I ask the class, then does that mean that both Y's and Z's rationales are equally acceptable?

Everyone senses that it would be strange and wrong to question a person's description of their own race. But some people still feel like there's something problematic with Y's choice. Isn't Y really biracial? Isn't Y polluting the data? Well, I say, what do you mean by *really biracial*? If race and ethnicity are implemented as self-selection of data values, then isn't your *real* race whatever value that you pick? Self-description doesn't prescribe a particular rationale as being more acceptable than another. This makes the census exactly the same as the dating sites. It wasn't problematic there for someone to potentially select Middle Eastern and East Indian on one site, and then for that same person to select only East Indian or only Middle Eastern on another. In other words, I continue, if we are implementing race and ethnicity by means of self-selection of data values, we are instantiating a particular idea of race, one that is both fluid and personal, even as the data values that are available to be selected suggest certain characteristics (such as color, sociocultural heritage, and nationality) as foundational to this idea.

Occasionally, a student will object that fluid and personal data implementation seems very unscientific. I explain that we could, of course, implement race and ethnicity data in other ways, by introducing another form of data creation practice. Other forms of data creation may inscribe race differently, even if we retain exactly the same set of potential values. (In other words, we don't change the available choices that the census or a dating site offers, but we do change how those choices are assigned.) For instance, our race data would potentially be more consistent if, instead of self-identification, data values are selected by a panel of independent assessors. For this census, this actually used to be the case before 1970; data was recorded by census takers rather than self-reported. We could return to this kind of protocol. Maybe we could choose assessors with special qualifications—a panel of experts. Or the data could be crowdsourced, so that as many people as possible could assign this data, to provide objectivity through averaging. Of course, the assessors could also be restricted to people with a stake in the matter. In a dating site, for instance, other community members could generate this data. It could be a requirement after going on a date with someone to enter data about them. At this point, one of the students might involuntarily yelp "No!" when the reality of such a data implementation regime sinks in. The class begins to shuffle its feet. They are beginning to realize that I am making them uncomfortable on purpose. We could, I add, make the judgments of these assessors more consistent by giving them tools to facilitate proper data collection. For instance, we could develop charts and diagrams to identify skin tones and facial features associated with particular races. South Africa developed such tools under the apartheid system, in which an official committee would assign (or potentially reassign) citizens to one of four race categories (Bowker and Star 1999).

The class is squirming now. What I am saying is not what they intended! They were hoping to promote objectivity, not to create a dystopia. In this case, however, changing the method of data collection recasts the data entirely. What I mean here is not merely that an "independent data collector" for the census might select a different value than Y and Z would select for themselves. I mean that should the census create data according to the "objective judgments" of "neutral assessors," the concept of race as described by the census would be transformed into something completely different. When someone describes their own race, they are telling others how they see themselves, as informed by their own experience. That experience may cause the person to weigh certain factors over others (as when Y focuses on the experience of being seen as a Black person in society, while Z focuses on the experience of having one Black parent and one White parent). In contrast, when someone's race is described by another person, the determination of race is made by different criteria, using different assessment processes. Because the "neutral assessor" can only employ criteria based on physical manifestations rather than on lived experience, race becomes biological rather than social. This horrifies us, because of the long history of injustices perpetuated on the basis of supposed biological differences between "races."

The larger point here, I would remind my students, is that all data partakes of the circumstances of its creation, and every technical decision may have vast consequences. Even more importantly, the same technical decision—for instance, the decision to use self-identification or an independent assessor—may have vastly different consequences in different situations.

I've used similar ideas and examples in professional development courses as well as with university students, and the working professionals find this whole discussion much harder to take. They find it difficult to imagine a data management protocol that "works" without an easily digestible set of "best practices" that have universal application, no matter the context of data creation and use. Whenever I teach such courses, I am beset by well-meaning professionals asking me to tell them what to do, ideally presented as a set of simple rules. Often, in these environments, students are especially interested in removing humans from all aspects of data collection and processing. So when we talk about examples like race and ethnicity, they immediately ask about automated alternatives, such as DNA testing. Usually, they will preface their remark by allowing that the idea of using DNA data for a dating profile or the census is a troubling prospect from the perspective of privacy concerns. But if those matters were resolved, wouldn't a DNA analysis be the better choice for understanding race and ethnicity?

Let's think about this for a moment. Let's imagine that we have a very accurate means of determining the relative percentages of a person's ancestry that arise from specific groups of people (something that, as I am writing this, commercial DNA tests do not currently provide). That certainly tells us something; however, it once again completely transforms the construct of race that we are describing. If self-identification is an assessment of lived experience in a particular sociocultural environment, and external identification is an assessment of physical manifestations, then DNA analysis is something completely different: an assessment of genetic material as received across unknowable generations.[10]

To see how these different ideas of race are not equivalent, let's begin with the story of Livie, as presented in a 2017 television commercial for personal DNA testing with Ancestry.com. Livie, an actual customer, had always considered herself Hispanic. But Livie's DNA results showed more ancestral diversity than she was expecting. "I'm from all nations," Livie exults. Previously, Livie had selected Hispanic when asked to describe her race. But now, she merrily imparts, she selects Other.

On its own, Livie's story may seem innocuous. But it takes on a new light if we compare it with another example. In 2006, the Harvard professor and public intellectual Henry Louis Gates Jr. hosted a series on public television called *African American Lives*. On the show, Gates and a team of researchers investigated the family history of Black celebrities, including Oprah Winfrey, Whoopi Goldberg, and Quincy Jones. DNA profiles were included as part of the show's research. In one episode of the series, Gates also investigated his own heritage. Gates's DNA profile revealed that his family ancestry was at least half European. Did these DNA results mean that Gates should understand his own race differently, that he was "really" biracial or even White? (After all, that's what Livie's commercial would suggest.) Gates dismisses this possibility. Gates is descended from enslaved people; his family had always been considered Black; he had always experienced life as a Black man in the United States. None of this changed as a result of the DNA profile. Indeed, several years later, Gates was arrested for "breaking into" his own house after he returned from an overseas trip. This incident made national headlines as a particularly stark example of racial profiling in policing, that one of the most famous scholars in the land would be viewed as a potential thief—in his own neighborhood, on his own property.[11]

The Ancestry commercial articulates the same position as the students who ask about automation. It quietly argues that two very different methods of data implementation—self-identification and DNA testing—express the same concept of race and ethnicity. Further, the Ancestry commercial suggests that self-identification is prone to error, while the DNA test is not. Livie, in changing her self-identification to match the DNA analysis, is happy to accept this proposition. But the Ancestry argument is clearly specious for Gates, for whom the lived experience of being Black in the United States is in no way equivalent to the results of any DNA test.

One might, when assessing this kind of situation, attempt to balance Gates's skepticism against Livie's blithe acceptance, perhaps to read Livie's case as a "vote for" understanding DNA and self-identification as equivalent concepts, while reading Gates's case as a "vote against" understanding DNA and self-identification as equivalent concepts. Gates's arrest, however, provides a clear indication of why such thinking can be dangerous. In a social reality of racial profiling based on physical appearance, DNA test results are superfluous.[12]

It's much easier to slip into this kind of disconnection—in which different forms of data implementation are not semantically commensurable—with concepts like race, which, despite their social importance and ubiquity, are challenging to define, impossible to agree on, and ruthlessly dynamic. Still, as with the earlier example of height, any concept will experience some kind of local mutation when it is implemented as data, within the context of some sociotechnical arrangement. But just as that sociotechnical arrangement produces semantic variation across local boundaries, it also lessens variation within those boundaries. Definitions of race might befuddle us, and yet as we have seen through looking at the census and online dating sites, certain regularities of everyday implementation produce a limited stability. It's a tacit, tenuous, and fragile stability, but it applies nonetheless.

A Novelist Talking about Danish Racism

Even with concepts as generally unsettled as race, the local stability of data implementations can be hard to recognize for what it is: limited, particular, and constrained. But when our position changes, and we find ourselves within another sociotechnical arrangement, we may find ourselves struggling to understand what about our environment has shifted and why everything seems just slightly askew. That's what happened to me at Louisiana.

As I sat in the concert hall and listened to Claudia Rankine and Synne Rifbjerg, I realized that in all the paperwork that I had submitted during my first few months in Denmark—to obtain legal residency, to process my appointment at the University of Copenhagen, to get a citizen registration number and a bank account and all of the public and private enrollments that required so much personal information from me—I had not once been asked to provide my race or ethnicity. Country of citizenship, sometimes, but race—never. For an American, this was very strange!

Indeed, publicly available demographic information from Statistics Denmark tracks country of origin, citizenship, and parents' country of origin, but not race or ethnicity.[13] Curious, I checked a Danish online dating site—Dating.dk, supposedly the biggest service in Denmark—and there were no profile components related to race or ethnicity there, either.

Is race and ethnicity data absent from population data and dating profiles in Denmark because it is a materially inconsequential characteristic, like eye color (one of the characteristics that Dating.dk does ask about)?[14] One perspective comes from the Danish author Jonas Eika. In October 2019, two months after my visit to Louisiana, Eika won the prestigious Nordic Council literature prize. In his acceptance speech in Stockholm, Eika pointedly criticized the Danish prime minister, Mette Frederiksen, for continuing a policy to place refugee families in special detainment centers. Frederiksen's recently elected Social Democratic government, said Eika, were hypocrites. The left-leaning Social Democrats might say that "in Denmark, we are all equal," but the reality was, according to Eika, that, "in Denmark, we have state-sponsored racism" (Eika 2019; DR 2019).

Eika's speech was confrontational (Mette Frederiksen was in the audience) and controversial (acceptance speeches for the Nordic Council prize are not typically so political). News reports focused on these aspects. But when I read the full text of the speech, I found it especially striking in its direct invocation of race. The topic of refugees, certainly, has been a sensitive one in Denmark. In recent years, many refugees have been Syrian. But refugees and other immigrants from across the Middle East—Iraq, Turkey, Bosnia—have been an increasing presence in Denmark for 30 years. In my neighborhood in Copenhagen, there were many residents of apparent Middle Eastern descent, including many born in Denmark—some of whom are now adults with children of their own.

In the portion of Eika's speech that addressed the refugee detainment centers, it would be possible to read his discussion of *race* as oriented around nationality. In other words, potential data values for race would contrast other nationalities (e.g., Syria) with Danish nationality. This would align with the demographic data from Statistics Denmark, which tracks nationality of origin and citizenship. If one were reading just this part of Eika's speech—the part primarily reported in the news media—one might assume that anyone born and raised in Denmark, with legal residency, would therefore be Danish. In other words, young Danes of Middle Eastern descent would be just as Danish as young Danes of Icelandic descent or young Danes whose family line has been in Denmark since Viking times.

Toward the beginning of Eika's speech, however, he associates "nationalistic" racism in the Nordic countries with an additional component: Whiteness. Here, Eika implies that "nationality" is not just about one's country of origin. Instead, "nationality" is the conjunction of race (based on Whiteness and its opposites) *and* country of origin. In other words, being White and Danish is not the same as being non-White and Danish. Accordingly, Danish children of, for example, Icelandic parents are perceived as *Danes* while Danish children of, for example, Syrian parents are not. Whiteness, according to Eika, functions as an unacknowledged component of Danishness.

My point here is not so much to debate Eika's viewpoint but to see it as a form of social critique that could be directly extended into the realm of data implementation. Eika's analysis suggests that the demographic data collected by Statistics Denmark is incomplete. If that data accounts only for country of origin and current citizenship and not for Whiteness, that data omits an important thread running through the current social fabric.

When I first read Eika's speech, I nodded in recognition, just as I had done when I was listening to Claudia Rankine at Louisiana. After all, I had been talking about the implementation of race data in my university courses for years. But then I remembered sitting in that concert hall among so many other White people, feeling slightly out of phase, and I realized that the roles I had observed had become reversed. I was bringing my own understanding of Whiteness to bear, and that understanding of Whiteness was deeply integrated with the local data implementations with which I was most familiar as an American. In contrast, Eika's remarks posit a different form of Whiteness, as it manifests in conjunction with Danishness (and in other Nordic nationalities). Eika situates this Nordic Whiteness within a specific legacy of past colonialism, which has found present expression in rising Islamophobia. This historical trajectory of Nordic Whiteness is, of course, very different from the Whiteness that obtains in the United

States, which takes shape against the enslavement of Black people and the displacement of Indigenous communities. If I'm restating the obvious here, it's only to emphasize that the technical implementation of Whiteness as data—of all data—is irreducibly local. It's also to emphasize how easy it is to lose sight of that local dimension, even with constructs that we struggle to define in a general way, like race and ethnicity.

When I think back to that dialogue from Danish class with Jonas and Gertrud, I remember again how natural it was to situate their conversation within my personal sense of locality, and how unsettling it was to realize that not only the fictional Jonas and Gertrud but everyone else in my class was looking outward from a slightly different basis of understanding. That day in class, I hid my sense of disorientation, not wanting to admit that I had been so oblivious to where I was, so narrow-minded and, well, stereotypically American. But that instinct was perhaps the wrong one. The thrill of travel arises from discovering the unfamiliar; how unfortunate to feel embarrassment in what should be a pleasure. A traveler who finds everything to be perfectly in order, just as expected, is missing something about the whole enterprise. Why go anywhere if you remain the same person everywhere you go?

Why should we expect anything different from traveling data? If our race and our height and who knows what else changes as we move from one environment to another—if our very understanding of race and height alters as a result of local data implementation decisions—this is not an embarrassing problem to hide, but a story to be shared.

REFLECTION: HOW IS DATA SITUATED?

Constraint and Creativity in Cognitive Categorization

As human beings, we have certain cognitive regularities in the way that we perceive and categorize external phenomena. Imagine that you're sitting at the breakfast table, just about to eat your oatmeal. Suddenly I grab your eating implement and ask "What is this?" "A spoon," you would almost certainly respond. Any quizzical looks or hesitation on your part would result from the bizarre circumstances of my asking something so obvious. Of course it's a spoon. What else would it be? And yet there are, of course, many correct responses, ranging up and down levels of abstraction. You could have said "cutlery" or "a utensil" or "a teaspoon" or "a porridge spoon" or a host of other answers that would either be more general or more specific than "spoon." But you probably never considered any of these alternate possibilities, because "spoon" came into your head automatically.

On the other hand, if I accosted you in the kitchen as you were stirring the oatmeal on the stove, and I grabbed *that* implement out of your hand and asked "What is this?" well, then your answer is not as predictable. You might say "mixing spoon," or you might say "wooden spoon" or "plastic spoon," depending on the material of your actual implement, or you might well say "utensil" in this case. You might also say "spoon"—but it's more likely that you won't, because the spoon that you're using to stir your porridge in its pot probably deviates from your run-of-the-mill, average *spoon* spoon in a variety of respects. It's likely to be bigger, constructed of a different material, and generally used for a different set of purposes. The stirring implement is certainly a spoon, of course, but it's not as . . . spoony? an example. Linguistically, we need to qualify it when we refer to it.

The situation with the two spoons demonstrates what cognitive scientists call *prototype effects* (Lakoff 1987). The significance of prototype effects is that categories in the realm of human cognition transcend simple logic. In a logic-based framework, category membership is determined by a set of necessary and sufficient conditions. If conditions A, B, and C hold, then object X is a spoon. In this conception of category structure, all spoons are equal category members: there are no spoons that are better examples of spoons than other spoons, because everything that is a spoon has fulfilled the conditions for membership in that category. If our brains really processed categories in this way, categories like *spoon* would be stable and universal. We could envision a form of logical proof that would conclusively establish whether something is a spoon, and both the stirring implement and the eating implement would be equally spoony.

But life, as humans experience it, is complicated and dynamic and demands more than logic to comprehend it. Accordingly, human perception of categories does not actually operate according to necessary and sufficient conditions. In real, everyday life, we look at the implement that we use to eat oatmeal and think "spoon," and through this lack of adornment indicate that our plain old oatmeal spoon is actually an excellent example of a spoon— very, very spoony. In contrast, some of us look at the implement that we use to stir the oatmeal on the stove and think "mixing spoon," by which additional clarification we indicate that the stirring spoon is not quite as excellent an example of spooniness. But others of us look at the stirring implement and think "wooden spoon" or "utensil" or "cooking spoon"—not "mixing spoon." With this utterly normal disagreement, we collectively indicate that human cognition does not treat all category members the same. All spoons are not equal in their spooniness.

A *prototype* is the most excellent of category members, the conceptual standard against which actual things that might be spoons are assessed. Unlike a methodical assessment of necessary and sufficient conditions, however, our use of prototypes to enact a graded structure upon our conceptual categories—better spoons, worse spoons—is usually automatic and unconscious. We don't look at the oatmeal stirring implement and consciously compare it to the oatmeal eating implement. We just think that one is a *something* spoon (a mixing spoon, a wooden spoon) while the other is merely a spoon. When pressed, we might be able to articulate some rationale for our designation. But we might struggle or even fail to do so, because the cognitive process that we employ here makes use of unconscious cognitive aids: habits, assumptions, sociocultural patterns, and layers and layers of unexamined regularities that we rely on to let us act quickly and spend less time thinking.[15]

The notion of prototypes is useful because it describes empirical phenomena that we all encounter. But it is also somewhat frustrating, because, unlike the logical certainty of necessary and sufficient conditions, it is inexact and unpredictable in how it manifests. Sometimes, for instance, category prototypes (the "spooniest" spoon, the "reddest" red, the most "father-like" of fathers) are based on the most typical or representative characteristics of a category. This seems to be the case with spoons: the spoon that we are most likely to eat with is the most typical spoon, and this is the best example. But category prototypes can also align themselves with ideals, or stereotypes, or other forms of distinction. A prototypical father tends to exhibit stereotyped characteristics, for instance—works outside the home, plays sports with the children, takes out the trash—rather than typical or ideal characteristics. Moreover, prototypes might be associated with a variety of contributing factors, some biological (how our brains and bodies work), some social, some linguistic, some cultural. Some prototype effects appear consistently across most people. There is a lot of regularity in how humans identify certain basic colors, for instance, and we are likely to select the same shade as "reddest" no matter the society that we live in or the language that we speak. But most prototype effects are of narrower scope. They might apply for people who speak the same language or people who were children during the same time period or, in some cases, even smaller groups, such as people who attended the same university or people who perform the same job function. Moreover, even in the best of circumstances, when something in the world (that spoon you're using to eat your oatmeal right now) aligns with the prevailing prototype quite closely (it's a totally average spoon that you'd use to eat oatmeal or cereal), our agreement on its category status will only be generally consistent and not completely so. Once in a while, someone will look at that spoon and say that it's a soup spoon or a coffee spoon or something else that's not quite *just* a spoon, because it's just a smidge too big or too small or too round or too oval or some such—and, to the extent that they can surface it, their reasoning will be defensible, because that's the way it is with these sorts of judgments. They are not logical entailments. They are relative to circumstances.

Local Conditions and Cognitive Prototypes

When I teach about prototype effects, I use interactive demonstrations to illustrate this push-and-pull of cognitive flexibility and cognitive constraint in different ways. I might, for instance, hold up an array of objects (like a bunch of different spoons) and ask students to write down what each one is; we then compare answers, and see where there is substantial agreement and where there is less agreement. These classroom demonstrations are informal approximations of the kinds of experiments that researchers in psychology, linguistics, and cognitive science have used to understand and describe different kinds of prototype effects. I've done these demonstrations many times, with different student populations. There is always some degree of variation from class to class, but with physical objects (like spoons) the parameters of variation are relatively constrained, and they tend to appear as quirks of individual perception rather than the consequences of different group memberships.

Still, sometimes local conditions surprise me. Another demonstration that I often do is to display a series of color samples and ask "What is this color?" (For convenience, I use items of clothing as my "color chips"—typically shirts, because I have a lot of plain, solid-color shirts that look vaguely rectangular when I hold them up.) Color categorization is particularly stable across cultures; we tend to agree on the best examples of fundamental color categories (e.g., we agree on the best red and the best yellow, whereas we disagree on shades that are farther away from those so-called focal colors). When I perform the color demonstration, I'll often begin with shirts that approximate focal hues (where everyone agrees). I'll usually include a lot of variations of one color, such as many shades of blue, and mix in colors that generate a lot of uncertainty like mauve (or is it fuchsia or magenta or purple or something else?) or mustard (or is it gold or brown or yellow or "baby poop," as a student once said). Once in performing this demo, I held up a red shirt and was surprised to hear a scattering of responses: red-orange, tomato, orange-red, even carrot, in addition to red. Huh, I thought. It was a red shirt. I was expecting everyone to agree with that. In previous demonstrations, that shirt had been unanimously declared as red. However, when I took a closer look at my own red shirt, I realized that it had indeed taken on an orangish hue. Well, I'll be, I thought. I had gotten used to my shirt being red, and I hadn't noticed that it had faded over time.

In a scientific experiment, like those on which my classroom demonstrations are based, the episode with my formerly red shirt would be considered a kind of error that compromised the results. That's because those kinds of experiments were designed to inform upon human cognition generally: how we identify color rather than how we describe specific material objects. These aren't actually experiments about *data collection*. But my demonstrations show how these domains—human cognition and data collection—are, in fact, tightly integrated. Even in the highly controlled conditions of a scientific experiment, people are describing the color of some actual thing: they are, in effect, generating data about a specific object just as much as they are assigning an abstract *hue* to an equally abstract *color category*. The experimental setup is an attempt to transcend those specificities. But in my demonstration, not so rigidly controlled, the specificities resurface. Of course the color of my shirt changes as my shirt ages. Of course data collection is specific to a time, a place, a collecting agent (human or machine), and other elements of context. Of course I, living slowly with my shirt, don't notice that it has changed color. The episode with my shirt is a classic, mundane, utterly typical, revelatory example of actual data collection.

It's interesting, then, that the most common reaction from students upon reading about prototype effects and other empirical realities of human categorization processes, is puzzlement about its utility. I think the cause is that students are thinking like data designers, and they are looking to ideas about human cognition to inform the design of better data structures. They imagine a design choice between developing a data structure based around prototype effects or developing a data structure based around logic-based rules. What the science of prototype effects illustrates, however, is not how data structures *should* be designed but how data structures *will* be implemented in real, live situations of data collection. In other words, no matter what kinds of data structures we attempt to design, when we use those structures to create data for actual things—when we describe the color of an actual shirt, like my formerly red shirt—we are making decisions using prototypical reasoning rather than pure logic. Perhaps we have designed a beautiful taxonomic structure, using all the proper rules of taxonomic hierarchy, as I describe in chapter 5. Our taxonomy might, indeed, be logic based. But when we use that taxonomy to describe things, those decisions—that data—will apply the taxonomic classes according to prototypical reasoning. This is one way, for instance, of understanding my height measurement story from the adventure essay in this chapter. A prototypical measurement in feet and inches is in whole integers, not fractions of an inch.

The integration between cognitive categorization and systematic data collection is, moreover, bidirectional. Another episode from a color demonstration provides an illustration. This time, I held up a light blue shirt and expected to hear a few different answers: light blue, baby blue, pastel blue, sky blue. Instead, the entire classroom thundered "Carolina blue!" I had used the sky blue shirt in the past, but not since I had moved to the University of North Carolina (UNC). *Carolina blue* is the distinct hue adopted by the UNC at Chapel Hill as one of its school colors (the other is white, but it is not a distinctive shade of white—there is no "Carolina white" to match the blue). In the United States, most universities have associated colors as an artifact of school-sponsored athletic teams. It is, however, unusual for there to be such widespread community agreement about the precise shade of the color, and to associate that shade with a unique name. Why is Carolina blue so locally salient? There are probably a few contributing factors, which include the need to distinguish Carolina's light blue from the dark blue of its neighboring rival, Duke, as well as local folklore associating Carolina blue with the springtime sky in Chapel Hill. More significantly, the local salience of Carolina blue has stretched beyond its original function as a university color used for official purposes. The shirt that I displayed in the demonstration was just a blue shirt that I had owned for years, since before I moved to the state. It had no logos or text that connected it to UNC. The students were UNC students; that was enough. We were also in a UNC class, of course, but honestly, I think the students could have happened upon a shirt of that hue in a boutique in downtown Durham and thought "Carolina blue." Further, even if (in legend) Carolina blue was inspired by the local sky, now that relationship goes both ways. It would be utterly ordinary for anyone lunching on the perpetually crowded patio of the Weaver Street Market, a few miles from the university campus, to look up at a cloudless April sky and think "Ah, Carolina blue."

Historical Efforts to Standardize Data Collection across Locations

This kind of reciprocal co-construction between defined, standardized data structures and the perception of phenomena in the world, outside of the context of data collection, is well documented in science and technology studies. Lorraine Daston (2015) describes a particularly evocative example of this in nineteenth-century cloud classification.[16] In seeking to aggregate meteorological observations from around the world, scientists faced two related challenges: first, to develop a standardized, universal language—a cloud classification—to describe cloud formations consistently, and second, to cultivate new habits of observation so that data collectors around the world would describe their local phenomena in the same global way. Prior to the development of the first international cloud atlas, as the standardized classification was called, Daston notes that various languages had evolved terms to describe similar cloud formations, such as mackerel sky in English, *ciel pommelé* in French, and *cielo empedrado* in Spanish, all of which refer to a cloud mass that looks like, variously, fish scales or a spotted animal pelt or cobblestones. The translation effort among languages was manifold, because it is not merely a case of the term *mackerel sky* being perhaps equivalent to *cielo empedrado* but of a mackerel sky cloud formation in London being similar to a *cielo empedrado* cloud formation in Madrid. If you do a modern-day informal test and use these terms as search queries in Google Images, the images are slightly different (see figure 7.1).

The mackerel sky images tend to appear more dense with little cloud scales than the looser dappling in *ciel pommelé*, for instance. But is that because clouds in France are generally a little looser than clouds in England—that

Locality 217

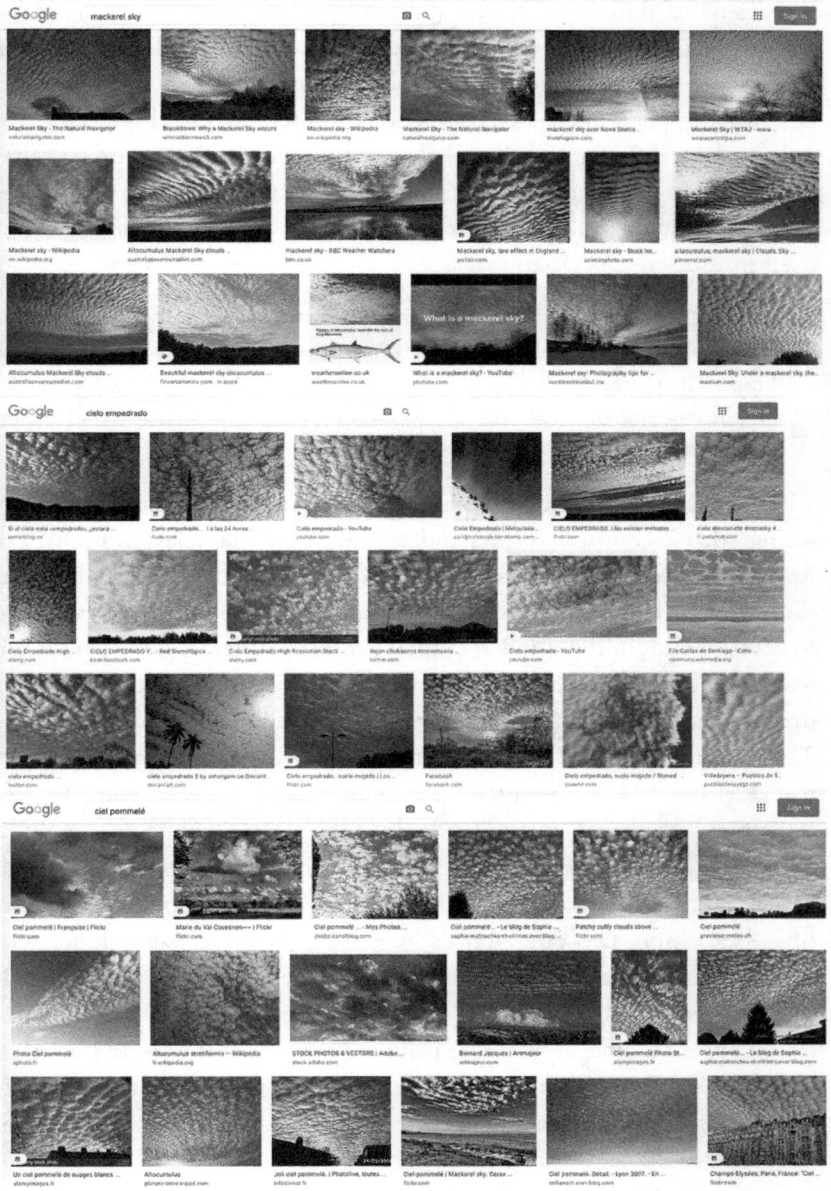

Figure 7.1
Image search results for mackerel sky (top), *cielo empedrado* (center) and *ciel pommelé* (bottom) show some small variation in the cloud formations: mackerel sky has the tightest formation.

is, because of differences in the phenomena themselves? Or is it because mackerel sky and *ciel pommelé* are describing slightly different kinds of phenomena—that is, differences in the categories (i.e., data values) used to describe the phenomena? Or is it because observers use those categories differently in different locations—that is, differences in how the data values are applied to the phenomena? It's because of all these different possibilities that developing a standardized set of categories for cloud formations is in itself insufficient. As Daston says

> Observers had to learn to see the sky in the same way, to divide up the continuum of cloud forms at the same points to connect the same words to the same things. Their attention had to be sharpened for the telling detail and blunted for the idiosyncratic one. Descriptions of cloud types functioned as templates and frames for observation. (Daston 2015, 52)

There are several points to make about this new, standardized "way of seeing" that accompanied the standardized cloud atlas (classification). The first is that, even if cloud observations reorient around the clearly bounded categories as defined in the cloud atlas, the application of those categories to actual clouds in the sky will exhibit prototype effects: some clouds will seem clearly altocumulus, while others will seem less altocumulus-y and may be designated as subsidiary varieties or even as other primary types. The second is that, as Daston documents, over time the local variation of mackerel sky and *cielo empedrado* begins to emerge anyway, with the new categories introduced by the cloud atlas. (That is, we will begin to wonder again whether an altocumulus in London is the same as an altocumulus in Lyon.) Third, the cloud atlas produces effects on perception beyond scientific observation. Daston describes how artists begin to paint clouds with greater attention to their structural details and differences, as they also train their eyes to see the distinctions defined by the cloud atlas. The way that clouds appear in images (whether painted or drawn or framed in photography and film) then contributes back into our observations of actual clouds, just as Carolina blue becomes a way to see the sky in Chapel Hill, in an ongoing process of mutual construction.

Daston's story of cloud classification tracks closely to Bowker and Star's (1999) influential discussion of the International Classification of Diseases (ICD). Bowker and Star positioned classification schemes and other data standards as a form of *infrastructure*. Similar to physical infrastructure like roads and bridges, the conceptual infrastructure of data standards facilitates and coordinates our activities by constraining them. More of us can travel faster and more efficiently on roads than bare terrain, but we can only go where roads have been constructed, and we need to obey traffic rules, and so on; and if the roads don't make space for bicycles and pedestrians, we are less likely to walk and bike, and to see walking and biking as viable transportation options. Classification schemes and other data standards shape our activities just as roads do, facilitating and constraining in equal measure. The ICD, which standardizes causes of death, facilitates the worldwide aggregation of death statistics and the tracking of global epidemics, just as the cloud atlas facilitates meteorological observations. In conjunction with standardized death certificates (which determine the possible number of concurrent causes and the relations between them, for instance), the ICD enforces certain ways of describing biological processes (e.g., death must arise from some kind of *disease*; old age is thus not a cause of death).

Indeed, this kind of story has become exceedingly familiar to anthropologists, sociologists, and historians.[17] Anyone who studies how data gets made understands that filling out a death certificate or recording cloud observations are similar processes to deciding on the color of a sky blue—light blue? Carolina blue? periwinkle?—shirt. There is no tenable distinction between everyday description (like deciding that your oatmeal stirring implement is a "mixing spoon" or a "wooden spoon" or a "utensil") and systematic, scientific description (like deciding whether cause of death is an underlying condition [such as chronic obstructive pulmonary disease, J44.9] or a proximate cause [such as viral pneumonia, J12.9] or a suspected but unconfirmed intermediate cause [such as COVID-19, U07.1] or all three or some subset).[18] Nonetheless, these absolutely normal, well understood, widely documented processes of locally influenced interpretive flexibility often seem to surprise data creators, aggregators, and users.

Data Creation as an Assertion of Values (An Extended Example from Library Cataloging Data)

It's a weird and sticky dissonance. In my discipline of information science, where the practice of data collection and management is something that we do, as well as something that we observe and theorize, scholarship can sometimes appear fractured. In particular, research that focuses on the theory and design of data infrastructure (e.g., scholarship oriented toward developing classification schemes like the cloud atlas or the ICD) has long proceeded from a standpoint that *incorporates* interpretive flexibility. "Every classification scheme is a kind of argument" has been a common refrain in such research for the past 25 years.[19] This phrase encapsulates a position that mackerel sky and cielo empedrado and altocumulus are different human ideas that capture different slices of experience, and applying any of these ideas to a cloud formation is like writing a sentence about the sky (with the idea that many sentences can be written to express the sky in different ways) rather than solving an equation about the sky (with the idea that there is one correct answer to discover the truth of the sky).

At the same time, research that focuses on best practices for implementing data with standardized infrastructure (e.g., scholarship oriented toward creating data using a classification scheme like the cloud atlas) seeks to *minimize* interpretive flexibility.[20] For instance, data quality criteria that emerge out of this branch of scholarship tend toward notions like accuracy, comprehensiveness, and consistency—characteristics that promote an idea of *good data* as the output of a mechanized process, more like an solving an equation than writing a sentence.[21] In the absence of additional quality criteria, accurate, comprehensive, and consistent data is unitary and fixed in scope: it is the same in London as in Lyon, the same in the nineteenth century as in the twenty-first century, the same in recording meteorological observations for short-term local weather prediction as in recording observations for longitudinal climate modeling. Of course, one understands the appeal of this position. Globally accurate, comprehensive, and consistent data can be frictionlessly aggregated and processed. If the human interpretive flexibility that leads to the appearance of prototype effects in categorization gets in the way of that goal, the appropriate response is to refine our practices as best we are able, to eliminate and reduce that flexibility as much as possible. Just because our efforts will have a limit doesn't mean that we shouldn't try to approach that limit as closely as we can.

There are several problems with this apparently reasonable proposition. The first involves its feasibility. Interpretive flexibility can be reduced—data infrastructure like the ICD and standardized death certificates does constrain it (I address this in chapter 4). The work of Bowker and Star, Daston, and oodles of others demonstrates both the limitations of these constraints and the immense effort that needs to be constantly applied in order to maintain the integrity of such constraints, even in their imperfection. The extent of this maintenance and repair work is almost always underestimated. But the second problem involves its desirability in the first place. The human interpretive flexibility that leads to category prototype effects is a cognitive advantage, not a deficiency. Sometimes it leads to miscommunication, misunderstanding, and confusion, for sure. But it also enables us to dynamically adapt to fluctuating circumstances, to evolve our understanding of the world around us, and to surface and negotiate disagreement.

To make this objection concrete, I'll use an example from library catalog data. Library catalog data may seem old-fashioned, but it remains a particularly useful source of data examples. It is excessively standardized, with an array of associated data infrastructure, including data creation protocols, controlled vocabularies, and file formats. It is produced by professional data creators who have completed specialized professional education. Even when developed for local use, it is created with an eye toward global aggregation—it is expected that library catalog records will be shared amongst institutions and merged into union datasets. Every catalog includes old data from many years ago and new data that has just been added. The data compiled for each record includes elements that appear conceptually complex (such as the subject of a resource) and elements that appear conceptually simple (such as the title of a resource).

My example involves the subject headings for the book *Protocols of the Elders of Zion*, a notorious anti-Semitic text.[22] The book, which first appeared in Russia in the early years of the twentieth century, purports to be an account of clandestine lectures in which Jewish leaders plot to overthrow European governments, presumably while laughing maniacally, twirling their mustaches, and making finger tents. The book is a hoax, created to deceive. It was produced to sow distrust about Jewish loyalties and motives. Despite its cartoonish nature, the *Protocols* was widely translated and found a worldwide audience. In the United States, automobile magnate Henry Ford used the *Protocols of the Elders of Zion* as the basis for a series of anti-Semitic articles in a newspaper that he owned, the Dearborn *Independent*. The book is, therefore, an important historical document, and many university libraries include multiple versions in their collections, in a variety of editions and languages.[23]

As a general guideline, the data creation protocols for library catalog records, Resource Description and Access (RDA) give precedence to the self-presentation of a resource. Titles and authors, for example, are directly transcribed from the resource itself whenever possible, even if the form of name differs from other sources. Similarly, aboutness is typically determined via examination of the resource's title and table of contents—that is, how it describes itself. Another way of saying this is that for purposes of cataloging, a resource is *about* the topics that it contains and not *about* the author's position on those topics. It does not matter if the resource becomes known for reasons other than its included topics or if its significance is not what its authors intend. The cataloger does not attempt to evaluate whether a book's position is justified, whether by evidence or by ethics. Assessment of worth is the reader's purview. By implementing aboutness data in this way—as the resource's main topics, established via the resource itself—the cataloger's work is conceptualized as primarily logical, rather than interpretive. The cataloger is systematically and carefully documenting what appears in a book without making judgments about accuracy, significance, or value.

In practice, when an author takes a position that goes against current consensus, the cataloger may generate aboutness data in a way that minimizes this conflict. For instance, take the example of *Raw Can Cure Cancer*, by Janette Murray-Wakelin. The book suggests that a raw diet can cure cancer, a position not currently endorsed via scientific consensus. The author's position appears to arise from personal experience: she credits a raw diet with her own survival of breast cancer. This book is associated with the following subject headings in the Library of Congress catalog:

- Murray-Wakelin, Janette (because it is a first-person account, the author is also the subject of the book).
- Breast—Cancer—Patients—Biography.
- Raw foods—Therapeutic use.

Here, two of the three subject headings focus on the author's autobiographical account of cancer survival, rather than claims for dietary cures. The subject headings do not repeat the "cure cancer" claim of the title; they do not associate raw foods with cancer treatment at all. Instead, the subject headings settle on the more general "therapeutic use" of raw foods—generally related to the treatment of disease, perhaps, but not specifically a treatment for cancer. The aboutness data thus stakes out a very ambiguous position. The subject headings do not assert that the book's characterization of a raw diet as cancer-fighting is spurious (for example, by including a heading for "quackery"). Indeed, the association of therapeutically administered raw foods and a breast cancer patient's survival is tacitly accepted. On the other hand, the selected headings do not present the book as *about* cancer treatment.

The case of *Raw Can Cure Cancer* can be interpreted in the way of asking "What is it?" about the implement used to stir a pot of oatmeal. The object in question is not a prototypical spoon, and so we might differ in our selection of discriminatory characteristics—a mixing spoon, a wooden spoon—or we might increase the level of abstraction and just call it a utensil. *Raw Can Cure Cancer* isn't a prototypical example of a book *about* cancer. Where the author proposes, via the title, chapter headings, and other cues, that the book is primarily about dietary treatments (a nonstandard treatment, so somewhat like a "mixing" spoon), the cataloger proposes that the book is primarily about a cancer survivor's experience (focusing in the substance of the narrative, so somewhat like a "wooden" spoon). Regarding the topic of cancer treatment through diet, the cataloger expresses this in the more general terms of therapeutic uses of raw foods, increasing the level of abstraction (similar to saying "utensil" instead of "spoon").

If we read the aboutness data this way, then the cataloger is making a mild assessment of the book's evidentiary basis, even if the subject headings are not actively questioning the book's accuracy. The book is not, in the cataloger's determination, an excellent example of "cancer treatment." Like all manifestations of prototype effects, this assessment might be entirely unconscious. The cataloger may understand this set of subject headings as merely the topics that the book contains, as established via the standard, logical, objective process of aboutness determination set forth in the data creation protocols. In other words, even as the data exhibits prototype effects, its interpretive aspects are easy to overlook. These subject headings arise from human judgment, but they appear otherwise.

The *Protocols of the Elders of Zion* constitutes a more extreme example. Purposefully, the data creation protocols of cataloging do not include provisions for intentionally deceptive forgeries. (Once again, it is not conceived as the job of the cataloger to make such assessments.) Accordingly, if the cataloger adheres to the data creation rules, any recategorization response to the book's self-presentation of its subject matter will be a relatively slight adjustment, on the order of *Raw Can Cure Cancer:* not a good example of "cancer treatment" but a better example of "the author's cancer treatment."

Because there are many editions and variations of the *Protocols of the Elders of Zion*, we can observe multiple examples of aboutness data for this work.

Table 7.3
Subject headings assigned to different manifestations of the *Protocols of the Elders of Zion* in library catalog records

Aboutness data #1	Aboutness data #2	Aboutness data #3	Aboutness data #4	Aboutness data #5
• Jews (Publication date: 1919)	• Jews—Politics and government (Publication date: 1934)	• Jews—Legal status, laws, etc. • Jews • Antisemitism (Publication date: 1970; reprint of 1934 edition)	• Antisemitism (Publication date: 2009)	• Antisemitism—History—20th century • Antisemitism—History—21st century. • Jews—Politics and government (Publication date: 2011)

Table 7.3 shows five such sets of aboutness data, taken from the online catalogs of the US Library of Congress and the University of North Carolina at Chapel Hill. (These examples were selected to illustrate a representative range of available data.)

Examples 1 and 2 make no reference to the text's position on its contents. The data asserts that the book's subject matter concerns Jews (example 1), or that it concerns Jews and politics (example 2).[24] (With such assigned data, someone entering the library to seek basic information regarding Jews might retrieve the *Protocols*.) In examples 3, 4, and 5, however, the subject headings incorporate the notion of anti-Semitism, either to complement (examples 3 and 5) or to replace previous data on contained topics (example 4). These references to anti-Semitism are not topical; the book is not about anti-Semitism. Instead, this new data refers to the anti-Semitic perspective taken on the contained topics (examples 3 and 4) or to the role that the *Protocols* has played within the history of anti-Semitism (example 5).

Examples 1 and 2 exhibit the kind of aboutness data that arises when the data creation protocols, intended to limit interpretive flexibility in this data, are followed. This is exactly the sort of data that we can see in the preceding example of *Raw Can Cure Cancer*, in which the book's claims for cancer treatment are generalized to "Raw foods—Therapeutic uses." Here, the subject headings do not associate Jews with conspiracy or treason, in alignment with the book's claims. Instead, the aboutness data is ambiguous.

The *Protocols'* discussion of Jews and politics appears anodyne, in no way different from any other book that might contain this topic.

In contrast, the explicit assessments of anti-Semitism that appear in examples 3, 4, and 5 are uncommon in cataloging data. The analogy here would be in adding a Quackery heading to *Raw Can Cure Cancer*—typically, the responsibility is on the reader to make such an evaluation. In the case of the *Protocols*, both its thorough debunking and its notoriety probably had to do with the unusual changes to its data implementation. Mere criticism, whether on grounds of racism or accuracy, is usually insufficient to change the catalog's approach to aboutness. For instance, we can look at the 1994 book *The Bell Curve: Intelligence and Class Structure in American Life*. This book suggested that intelligence (as measured by IQ tests) predicts life outcomes more reliably than social, economic, racial, and other factors. The book was widely criticized as both perpetuating racial stereotypes and misinterpreting its source data. The subject headings for this book take the typical, *Raw Can Cure Cancer* approach, in which its racist implications are ignored rather than repudiated:

- Intellect
- Nature and nurture
- Intelligence levels—social aspects
- Educational psychology

This situation demonstrates that, exactly as we would expect from the vast trove of empirical evidence we have about data collection practices, attempts to generate aboutness data as a kind of equation to solve—identifying and extracting the topics that a resource contains—are limited in their power. Catalogers modulate what books claim to be about all the time, playing down controversies by omitting or generalizing them. This is human interpretive flexibility in action.

Cherishing the Humanity in Our Data

Best practices for data implementation suggest that data quality suffers from these ghastly actions, in which human data creators act like humans. But I disagree. In the *Protocols* example, the data that plays closest to the rules (and reaches toward the static form of an equation) is *less* valuable than the data that openly transgresses the rules (and so approaches the flexible form of a sentence). Of all the aboutness data in these cataloging examples, the assertion that the *Protocols* plays a significant role in the history of anti-Semitism is the most important. This bit of data both situates the work's significance and is impossible to know from a putatively objective inspection of the book itself. Equally important, from my perspective, is the assertion by an individual cataloger, as a representative of both a particular library and the institution of librarianship, that the *Protocols* is anti-Semitic. This is not an assertion of fact; it is an assertion of values. We believe, the data says, that it is important to designate this book as anti-Semitic, even as library aboutness data does not generally include such information. In other words, we (the cataloger, the library, the institution of librarianship) are not just saying that "some people think this book is anti-Semitic" or "scholarly consensus is that this book is anti-Semitic." *We* think this book is anti-Semitic. It's this conviction that makes this data especially meaningful. In contrast, the aboutness data for *The Bell Curve* and *Raw Can Cure Cancer* is accurate, consistent (it follows the data creation protocols), and reasonably comprehensive (it aggregates each work's main topics). But this data doesn't give us much that we couldn't get from automatic processing of the text—for example, using statistical properties to select some important words—*except* when we look at the topics that the cataloger has omitted or modulated from the book's self-presentation. Those tiny judgments tell us that the cataloger (and, by extension, the library and institution of librarianship) finds the *The Bell Curve*'s discussion of race and intelligence to be controversial and problematic and that the cataloger views *Raw Can Cure Cancer*'s claims about dietary treatments with some degree of skepticism.

These judgments are valuable because, in offering an interpretive frame, they both contextualize and question the other data elements. Indeed, such judgments assert that putatively objective topical data is an insufficient mechanism to characterize the aboutness of these works. To say merely that the *Protocols* is *about Jews* or that *The Bell Curve* is *about intelligence* is like saying that Harvey Weinstein—whose unmasking as an inveterate, extreme sexual harasser accelerated the #MeToo movement—was a movie producer. Can Weinstein's career in film be properly understood without expressing some judgment upon his behavior toward women, which occurred in the context of his professional activities? Can *The Bell Curve* be properly understood without expressing some judgment upon its racist implications, which arise in the context of its claims regarding the genetic basis of intelligence? Aboutness, here, *requires* assessment.[25]

Note that I don't need to *agree* with the cataloger's expressed judgment. Once again, it's like two people looking at an implement used to stir a pot of oatmeal. One person says that it's a wooden spoon, by which is meant that, in order to understand what's important about this spoon, you need to consider its material. One person says that it's a mixing spoon, by which is meant that, in order to understand what's important about this spoon, you need to consider its function. The important thing about these statements isn't whether the spoon is *really* a mixing spoon or a wooden spoon. The important thing about these statements is that each one asserts a different basis for the spoon's significance: material or purpose. Nor does it matter what I think when I look at the spoon, whether it's mixing spoon or wooden spoon or utensil. The important thing for me to consider is that someone else thinks this spoon is a wooden spoon and that its material is vital to it. How do I understand the spoon now, knowing that someone else has made such assertions about it?

My overall point is that, in their zeal to create consistent, stable data, library catalogers have likewise constrained the construct—aboutness—that they were generating data to reveal.[26] Maybe, to a cataloger in a medical library today, *Raw Can Cure Cancer*'s aboutness can be understood only through the lens of quackery. That's a valuable assertion. Maybe, to a cataloger in an anthropology library tomorrow, *Raw Can Cure Cancer*'s aboutness can be understood only through the lens of "dietism"—a social fixation on the importance of diet. That's also a valuable assertion. The value of such human judgments is indebted to their specificity, to being located in a particular set of circumstances as lived by a person. Two contrasting determinations of aboutness for a single work don't negate each other; they complement each other.[27]

This is an unorthodox stance on data quality. More typically, professional (human) data creators locate their value as creating *better* data than is possible with automated techniques: better in the sense of a direct comparison using similar quality criteria (that is, more accurate, more comprehensive, and more consistent). Certainly, a person will describe the topics that a book contains better than a computer can do, in terms of creating a set of data values that make sense to other humans in what humans understand *topics* to be. However, as the Cranfield tests showed in 1967, better data in this sense—compiling included topics in the manner of objective facts—does not necessarily lead to better information retrieval outcomes. Under such a lens, the value of human judgment in data generation is negligible.[28]

If performance on retrieval metrics is our goal, then it matters little that automated techniques do not actually engage with aboutness at all, that they manipulate proxy data rather than attempt to perform interpretive judgments. Amazon doesn't know what books or movies you actually like or why you like them; Amazon tracks data about your purchases and clicks that it uses as a proxy for the concept of your taste. A search engine doesn't know what any Web pages are actually about, and it doesn't know what kind of content will actually be of use to you in your circumstances; it tracks other data that it uses as proxies for the concepts of aboutness and for your unknowing (or bored) self. These proxies are acceptable inputs for automated techniques of sorting and ordering, and the results of these techniques appear to work acceptably well for many retrieval and classification tasks.[29]

Or so we are led to believe. But this is nonsense! The systems that rely on these proxies do not work well at all! They only work when we keep to a mechanical understanding of sorting and ordering rather than a human one. We've merely become accustomed to poor outcomes, quietly accepting results from information systems that we'd never accept from other people. For instance, as I was writing the first draft of this chapter in July 2020, I typed "racism truth" into the search engine Duck Duck Go, and the top three results suggested that systematic racism in the United States was a lie put forth by the Democratic Party (see figure 7.2).

The search engine's use of proxy data to rank a set of documents that match my query did not make an error. The search engine was operating

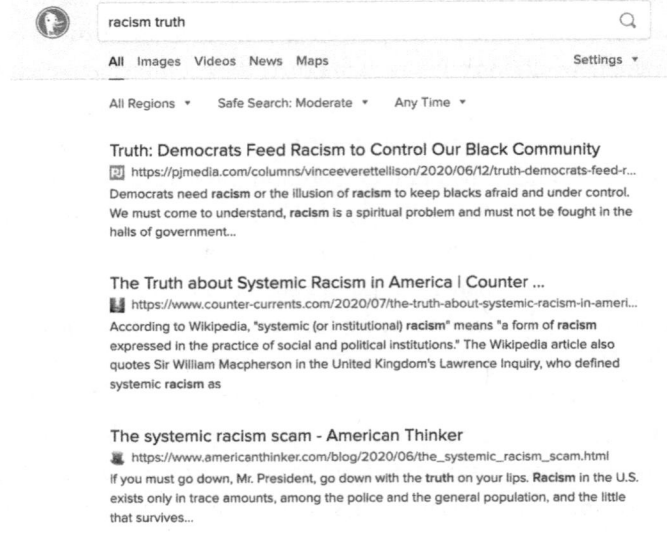

Figure 7.2
The top result for a July 2020 search on DuckDuckGo contended that systematic racism is an illusion put forth by the Democratic Party.

as designed: to approximate a constrained sense of the human concept of aboutness that has literally nothing to do with the interpretive judgments of actual humans.[30]

If we accept the regime of automatically extracted proxy data that produces such results, one reason is our longstanding conception of data quality, specifically our tendency to view human-generated data and machine-generated data in similar terms. When automated techniques did not exist, it seemed necessary to enroll a tremendous technical apparatus—data creation protocols and attendant technical formats and stabilization devices such as controlled vocabularies—to curtail human interpretive capacities. How else to facilitate data management and retrieval operations across large collections? This was the environment in which card catalogs and print indexes were developed, promulgated, and standardized, and the way that we understand data quality has remained pretty consistent. In short, we do our best to ensure that people, when they create data, do so as if they were actually machines instead of sentient beings with thoughts, feelings, and values. We prize humans as data creators because they are better machines than our machines are currently capable of being, but our idea of quality is machinic nonetheless.

What if, instead, we embrace the humanity (and attendant locality) in our data implementation, rather than attempting to eliminate that humanity? By this, I don't mean that we should eschew standardized data infrastructure like the cloud atlas, the International Classification of Diseases, or Library of Congress Subject Headings, with their accompanying creation protocols. Even if we view data as written like a sentence rather than solved like an equation, we still need technical infrastructure to make that sentence intelligible: codified vocabularies, usage rules, and generally agreed conventions.[31] (The taxonomic structures that I exalt in chapter 5 serve such a function.)

One way of thinking about this is to understand data creation in the same way that we understand any other communicative process that takes place within an established history of genre conventions and accompanying media technologies.[32] Genres are endlessly malleable and in constant motion, and yet they form a recognizable substrate upon which to interpret and appreciate individual works. We enjoy the detective novel or situation comedy that bends some conventions in the service of a particular story, even as adherence to other conventions provides a sense of familiarity and stability. But a work that follows genre conventions too closely? It's boring, formulaic.

If we understand data creation as communicative action along the lines of genre, we might find it easier to adapt data creation protocols to encompass and value a wider range of human interpretive flexibility. Via such an understanding of data implementation, we might, for instance, implement the notion of aboutness by making reference to different constituent components for different works, as determined by the person generating the data, in a particular moment and location—potentially to be changed by someone else when the moment changes and our sense of what's important evolves. Data creators are encouraged to act as human beings, rather than machines.

Of course, to some extent, this is what we already do—what we always have done. But instead of ignoring or denying the human in our data, we can surface it; instead of cleaning up or omitting the human, we can value it and see what it can teach us. Whether we are identifying the color of a T-shirt, describing cloud formations, selecting a cause of death, determining what a book is about, or associating a person with a race or ethnicity, we are people employing our values to make judgments. We're writing sentences about the phenomena that we're describing, not solving equations about them.

Oh, please, one might object. This seems terribly anachronistic. Algorithms, after all—like those employed by search engines and e-commerce systems—operate mathematically, on quantifiable data. They don't operate poetically on human data! If people say that a certain shirt is Carolina blue or sky blue or light blue, it doesn't matter; we can actually say what color a shirt is, precisely and quantitatively, in terms of CMYK values or RGB values, no words necessary.

Indeed. That's my point. Providing the RGB values for a shirt is a different kind of statement from saying it's Carolina blue. What's more, the RGB values are not as exact and invariable as we might naively imagine. Ask anyone who has ever carefully selected a particular color for the background of a slide presentation, precisely expressed with specific values for red, green, and blue. What a lovely blue it appears on your Mac PowerBook, as you design your presentation in your sunny office. Then, when you deliver your talk and the color is projected onto the screen, in the darkly lit conference room with no windows, that same color looks dull and sickly. It's the same RGB value, but it's not Carolina blue anymore. And isn't this a mundane and common experience? So yes; our algorithms follow the data that we give them. If we provide automatically generated, quantitative proxies for human judgments, we probably shouldn't be surprised when the decisions that the algorithms render don't seem very human, either. Instead, if we want our algorithms to be more humane, we should design them to incorporate the humanity in our data.

CONCLUSION: STILL LIFE WITH DATA

One day in the fall of 2008, I was roaming around the Elliott Bay bookstore in Seattle, blowing off steam. I'm not sure what made me pick up *Still Life with Oysters and Lemon*, by the poet Mark Doty—kismet?—but I found it captivating. Doty's book locates the sublime within the ordinary, in the form of seventeenth-century Dutch still life painting. Dutch still life paintings depict everyday items in familiar settings, with meticulous realism. By all rights, these works should be boring and repetitious. And yet, by attending to different characteristics in their portrayals, each artist produces a distinctive composition. Encountering the painting of the book's title, for instance, Doty marvels at the artist Jan Davidsz de Heem's "bravura lemon peel," which luxuriates in the lemon's textural qualities: its ridges and bumps (see figure C.1). De Heem's still life with oysters and lemon is both empirically accurate and decidedly singular, the product of one artist's vision and skill. When we contemplate Dutch still life as Doty does, we are not surprised when two lifelike depictions of similar objects evoke a different scene. Rather, we *expect* that a similar painting will find a different expression of the same source materials. And indeed, William Claesz Heda's rendering of oysters, lemon, and glassware focuses on the lemon peel's glinty, metallic hue, its textural qualities muted in comparison (see figure C.2).

My word! Doty's book encapsulated everything that I had wanted to say in my almost-finished dissertation! Description is a kind of art. Description interprets as it documents. As Doty says,

> A painting of asparagus, a painting of gooseberries, a painting of five shells arranged on a shelf. Exactitude, yes, but don't these images offer us more than a mirroring report on the world? What is it that such a clear-eyed vision of the particular wishes to convey? A way to live, perhaps; a point of view, a stance toward things. (Doty 2002, 47)

Figure C.1
Jan Davidsz de Heem's painting, which the Metropolitan Museum of Art now titles *Still Life with a Glass and Oysters*, includes an opulently bumpy lemon peel. (Metropolitan Museum of Art, New York. Purchase, 1871.)

Conclusion

Figure C.2
William Claesz Heda's *Still Life with Oysters, a Silver Tazza, and Glassware* depicts similar objects to de Heem's painting and yet to very different effect. (Metropolitan Museum of Art, New York. From the Collection of Rita and Frits Markus, Bequest of Rita Markus, 2005.)

Ever since, I've used *Still Life with Oysters and Lemon* as a little coda when I teach organizing information, to crystallize everything that came before. Students love Doty's beautiful language, but it's not always an easy leap to connect still life with data. Still lifes are certainly realistic descriptions of everyday things, but they are also paintings, a medium in which the creator has complete control over the canvas and from which we have become used to seeing the creator as an artist. In contrast, when we consider data collection as description, the creator is more constrained in what they can express and how they can express it. The creator may be limited to selecting from a controlled vocabulary of defined values, or limited to performing a measurement according to specified protocols, with automated equipment. Such conditions of practice seem quite different from that of the still-life painter.

To show how Doty's understanding of description can encompass data collection, I adapted one of Doty's own rhetorical strategies. Doty revealed memories of certain objects and how they become singular for him: his grandmother's peppermints, an enormous dish he bought with his late partner, Wally. So I tell my students about my great grandmother, whom we called Grandma.[1] When my sister and I were very young, we would sometimes spend an afternoon at Grandma's house. For lunch, Grandma always gave us open-face peanut butter and jelly sandwiches, cut into four equal squares, made on packaged white bread with Jif peanut butter and Knotts Berry Farm boysenberry jam. After the sandwich, we could each select one cookie from Grandma's tin. Grandma did not bake. Grandma never cooked anything from scratch! There were only ever packaged cookies in the tin, usually Nutter Butter or Chips Ahoy.

Lunch at Grandma's was merely a mundane arrangement of premade parts. But it was nonetheless distinctive and special. My sister and I both have extremely vivid and fond memories of it. Moreover, we had the same ingredients at our house too, but we could never have Grandma's lunch at home, even if we tried to replicate it. Somehow, Grandma had her own particular way of applying the peanut butter and the jam to the bread, and of determining the proper moment to retrieve the cookie tin from its place on the top of the refrigerator. Perhaps it wasn't bravura, but it was Grandma's boysenberry, just as much as it was de Heem's lemon peel.

I wasn't sure what to expect when I started relating this anecdote in class, but my students have seemed to understand what I was getting at. Doty calls description a "loving art," by which he means that when we apply interpretive judgment to describe something, we show how we care about it. And we do this in a public way, so that, in our care for the object being described, we also care for the audience of the description. In this way, description, in its sense of the particular, incorporates a form of intimacy, just like sharing a memory. Look at this! A description says. Isn't *this* aspect of *that* thing something to remark upon? Can you see it now, how marvelous *this* thing is in *that* way? My students can see Grandma's manipulation of prefabricated components in a similar fashion. The cookies don't need to be homemade to show care, when they are kept in the tin and allowed only at specified times. And the sandwiches are ordinary but also beyond replication—lots of American grandmas make, or made, peanut butter and jelly sandwiches, but none of them are the same. Extrapolating from my story about lunch at Grandma's, the students begin to assimilate what it means to see description as a loving art, from mundane data collection to painting a still life. For most, this seems like an exciting prospect. It's inspiring to find a form of art in selecting a data value from a list, and love in configuring a measuring device. We end the class feeling exhilarated.

Conclusion

However, when actual datasets come under scrutiny, it becomes more challenging to accept description as a loving art. I've seen this in my Metadata Architectures course, which I mention in chapter 2. In this class, students undertake a semester-long project where they first create detailed implementation guidelines for an existing data standard, and then they create a class dataset using their guidelines. (Their dataset describes video games, and to enable robust comparisons, everyone describes the same three games in addition to seven other games that they select.) Subsequently, they assess the dataset that they have created and use that assessment to inform a position paper in which they present their perspectives on data quality and how to achieve it.

I've conducted six different iterations of this project since 2014, at two different universities.[2] When I first conceived of the project, I thought it likely that the students would find more variation and ambiguity in their data than they expected, but I didn't know how this would manifest, or what they would make of it. It has, therefore, been interesting to observe a constant tendency in the position papers, in which students embrace *some* forms of data collection as a loving art but are equally adamant that *other* forms of data collection should not be seen in those terms. Specifically, students agree that it can be worthwhile when subjective forms of data—what we might call interpretation, or opinions—demonstrate interpretive diversity. But only harmful effects result when objective forms of data—in other words, simple description, or facts—demonstrate similar interpretive diversity.

As I touch upon in chapter 2, the biggest surprise for students in their project is that *all* the data they create is equally inconsistent and uncertain. Students anticipate that there will be differences in data elements that seem clearly interpretive to them, such as the Mood of a game, even though Mood uses a controlled vocabulary to constrain its values. But students don't anticipate that there will be differences in Price, Release Date, Language, and other data elements that seem more straightforwardly descriptive—as though they can be understood in only one way. The students are all conscientious, and they make very few errors in the data that they collect. But the data still varies. Often, this variation arises because the students are describing the *game* at different levels of abstraction; they aren't actually describing the same thing. (This phenomenon is discussed in chapter 3.) The Price and Language vary because one student is describing the original price for the version released in the United States on CD-ROM, while another student is describing the price for the version that one would access today via a streaming service. Although we discuss the challenges associated with functional equivalence in class, students don't realize the extent to which this will affect the data that they collect in actual practice. More significantly, they don't realize the extent to which this data variation will trouble them when they see it occur.

In the class, we perform a few joint data analysis activities where we all look at certain aspects of the dataset and compare our findings. When we look at elements like the Mood of a game during these exercises, students find it interesting when different data creators select different data values. We examine how these differences are often not cumulative but result from distinct variations in perspective. For example, if some students find a game to be Dark and other students find a game to be Mysterious, it doesn't necessarily mean that the game is both Dark and Mysterious. It could mean there are two, equally valid interpretations of the game's Mood: one Dark and the other Mysterious. Students are likely to see these situations as similar to the different lemon peels painted by Heda and de Heem, in which one artist finds the lemon peel to be glinty while the other finds it rough: not contradictions, just different stresses.

In these collaborative analysis exercises, we also look at elements like Price and Language. To me, there can also be value in this kind of data variation. When we see different prices for the game that we commonly identify with the name *Skyrim*, for example, we see that Skyrim is a thing with different versions, and we can trace and relate those versions to establish their prevalence and provenance. We may additionally identify patterns that are similar to the Mood element, in that we see several consistent, distinct interpretations where the majority of data clusters around X or Y. To me, that's also like the difference between Heda's lemon and de Heem's lemon, or my Grandma's peanut butter sandwiches and your grandma's peanut butter sandwiches. It's data as a loving art. There is a beauty to it.

For my students, however, the situation with Mood and Price is not the same at all. In their position papers, students often argue that interpretive diversity in Mood data is good, but interpretive diversity in Price data is not. Mood is subjective opinion; Price is objective fact. Mood is interpretive, but Price is not. In their papers, the students contend that a boundary must be maintained between subjective and objective, opinion and fact, interpretation and description. They make statements such as the following:

- "The more objective the element, the less interpretive diversity is allowed." (A student from spring 2015)
- "Interpretive flexibility should be limited for metadata that can reasonably be understood to have a defined answer or set of answers." (A student from fall 2017)

In their papers, students attempt to define the "more objective" elements, the data "that can reasonably be understood to have a defined answer or set of answers." For example, one student wrote

> Objectivity is all about numbers. Facts. Data that is infallible to a high degree. If I were to ask you who the first President of the United States was, there is little room to argue about who it may be. No matter how you "feel" about the first President, it will always be George Washington. It is just the same, what if I asked you about what consoles you can play Skyrim on? (Another student from fall 2017)

This kind of argument has two parts. First it asserts the existence of facts (objective description) in opposition to opinions (subjective interpretation). Then it places some elements of the student dataset within the realm of facts. This particular student from 2017 proposes that facts, unlike opinions, are accurate, invariable, and context-free. This student then contends that "the consoles that you can play Skyrim on" are facts.

Such arguments seem reasonable at first, but they ignore the empirical reality of the datasets that the students collected. For instance, in the datasets, the "the consoles that you can play Skyrim on" did vary. However, the data was not inaccurate. The data varied because the student data creators interpreted both "consoles" and "Skyrim" with different degrees of specificity. And yet those differences in judgment were perfectly reasonable; there are multiple, valid ways to interpret "consoles" and "Skyrim" as concepts. Similarly, let's look at the example that the student from 2017 offered as an obvious fact, that George Washington was the first president of the United States. Yes, as long as we understand "president," "United States," and "George Washington" in the same way, we are likely to agree that the first president of the United States was George Washington. But there are reasonable circumstances in which we might understand those concepts differently as well. Before the current US constitution, for instance, the United States was organized under the Articles of Confederation, which didn't have an executive branch in the same way that the current constitution establishes, but which did have a president of the Congress of the United States. Under the Articles of Confederation, the first president of Congress was Samuel Huntington. Does that mean that the first president of the United States was "really" Samuel Huntington? No. It means that there's just a little judgment involved when it comes to instantiating concepts, be they "president" or "United States" or "consoles" or "Skyrim" or grandma-made peanut-butter and jelly sandwiches. When it comes to "president" and "United States," the association is often automatic, at least for students at my American university. But the association between "consoles" and "Skyrim" also seemed automatic to my students. However, the variation that appeared in the datasets showed this assumption to be false.[3]

Variations of this argument appear in paper after paper, year after year. But none of these papers can account for the empirical reality of the data. As much as the students want facts to be different in kind from opinions, the data itself does not support this. None of the data elements are free from interpretive judgment, and none of the data elements are free from variation, even as all of the data elements are equally accurate, in that they are based in verifiable evidence. Ultimately, when these students argue that objective, descriptive, fact-based data is different in kind from subjective, interpretive, opinion-based data, their appeals are emotional rather than logical. Factual data represents data that we just *should* all see in the same way—we *should* agree on the Platform for Skyrim, or its Price—even when we demonstrably *don't* see this data in the same way. Indeed, facts emerge most strongly when this dissonance becomes uncomfortable. Something seems wrong with the world when we generate different data for Skyrim's Platform, even when we can understand the circumstances under which all that data is equally correct. And when data variation causes this kind of anxiety, then we have entered the realm of facts. Accordingly, over the years, as I've read variations of this position paper, I've come to see it as a kind of yearning. The students yearn for a persistent, logical differentiation between description and interpretation, even as no logical basis presents itself.

At first, this yearning for facts may seem like a direct response to conditions of public discourse in the US, when the forty-fifth president became notorious for constant lies.[4] But the data that the students are reacting to is not false, nor is it meant to deceive anyone. It is inconsistent not because it is incorrect, but because it varies in perspective, emphasis, or level of specificity. Moreover, the position papers that were written in 2017 and 2018 are essentially similar to the position papers that were written in 2015 and 2016, before anxiety over fake news, alternative facts, and misinformation began to escalate. Concerns about insufficient evidence, I think, provide only a partial explanation.

I tend to see the students' yearning for facts as more about accountability than truth. If we contemplate using data to inform a policy decision, it would be reassuring to believe that a factual dataset has more claim to universality than an artist's rendition in a still life. In contrast, if data is a loving art, and there is no clear distinction between description and interpretation, then any data involves more uncertainty and more risk than it may be comfortable to admit. There is no refuge from responsibility in being data driven, because data is more realistically understood in the mode of Shakespeare than in the mode of science, with our ideas always subject to revision.

The role of the data creator is especially frightening in this context. If data is a loving art, then data creators are always describing *their* world when they describe *the* world. Moreover, these two worlds can't really be separated. My Grandma, I'm pretty sure, just made her peanut-butter sandwiches the way that she made them. Open face, equal amounts of peanut butter and jam—that was just what a peanut-butter sandwich was. And as my Grandma made them, so do I, even today, with a single slice of bread. That just seems right to me. But when I first told that story in class, a student said, "Your Grandma made open-face peanut butter sandwiches?!" as if open-face peanut-butter sandwiches were a perversion of the natural order. Was the student biased? Was I biased? The word *bias* doesn't really make sense here, does it? If data is a loving art, there is no unbiased way of perceiving, or describing, a peanut-butter sandwich or anything else, manmade or natural. We all see with an artist's eye. If we are data creators, that is certainly a very scary situation! If we understand data as a loving art, then data creators set the conditions under which the world is understood. That's a tremendous power, and we should be wary of it. But a yearning to deny the art in data doesn't diminish that power.

Ulrika Kjellman (2017) describes a particular case of scientific data collection that makes this clear: research conducted by the Swedish Institute for Race Biology from 1922–1958. Kjellman divides the institute's activities into two eras, under two different directors: the first era is 1922–1935, and the second era is 1935–1958. In the first era of the institute, its researchers used a combination of physical measurements and photographic documentation to perform a data-driven derivation of races in Sweden, identifying three distinct racial types: Nordic, East Baltic (Finnish), and Lappish (Sami). Kjellman is particularly interested in the photographs. The lighting conditions, for instance, emphasize certain characteristics in the depicted subjects: that Nordic specimens were lighter-skinned and taller, whereas Lappish specimens were darker and smaller. In the institute's research, analysis of this data—objective, conventional elements of physical appearance, such as height and skin color—demonstrated the superiority of the Nordic type. Of course the analysis produced such findings; the assembled facts could not show otherwise.

In the second era, notions of racial superiority were de-emphasized, and data collection also changed. In the new data regime, "measurements and estimations of bodily characteristics were abandoned, as was the use of photography to document racial traits." The institute turned toward population genetics (internal data), rather than physiognomy (external data). The data collected during this second era, in Kjellman's analysis, was less ideologically motivated than the data collected during the first era. Still, in the second era, data from the institute endorsed certain social policies, such as enforced sterilization, that are condemned today as racist. Kjellman, therefore, labels the institute's research as pseudoscience:

> Even if race biology, at that time, was thoroughly institutionalized in the Swedish scientific domain, I do not consider it as a proper science, worthy of the appellation—the scientific practice of the institute was permeated by ideological assumptions and the methods put into practice highly biased by a racist agenda.

Certainly, the racial distinctions put forth by both eras of the institute would find few adherents amongst current scientists. Nonetheless, Kjellman's analysis shows that data from both eras is eminently factual, if we employ the same kinds of arguments put forth by my students. The Swedish researchers in the institute's first era may have used data for height and skin color to what we now see as odious effect. But height and skin color, as determined via measurement and photograph, are assuredly "infallible to a high degree" in the manner that my student from 2017 describes. Kjellman, in her stress on ideological assumptions, would likely object that the institute's data collection was unduly influenced by racist beliefs: in other words, that the selection of data elements was tainted by an unscientific

Conclusion

approach. I think, however, that this case is better understood as typical, rather than exceptional. In both eras, the institute's scientists collected perfectly conventional, quite unremarkable factual data, presumably with excellent standards of accuracy. In other words, the institute did not deviate from scientific standards; it followed them.

The Swedish Institute for Racial Biology does not seem repugnant today, I would suggest, because its data was insufficiently factual. It seems repugnant today because we disagree with its underlying values. But if data is a loving art—if all data emerges from a particular way of seeing—then all data is created with reference to some set of underlying values. That's why the Swedish Institute for Racial Biology doesn't escape our condemnation by obscuring its values with an appeal to facts. Accordingly, the lesson to be drawn here is not that data creators should attempt to excise their values from their data, but precisely the opposite. Data creators should attempt to surface and take ownership of their artistry and the values that inform it. Good data, in other words, is not value-free. Instead, good data allows us to more easily recognize—and potentially reject—the values that it expresses in its loving art.

Of course, for data creators—and I include here designers and collectors of new datasets, aggregators and cleaners of existing datasets, and anyone else who participates in ongoing processes of data management—such a directive is consternating. Owning one's artistry is a confounding task that resists both completion and perfection. (We'll always be surprised that not everyone makes their peanut-butter sandwiches like our grandmas did.) Owning one's artistry, therefore, means accepting only partial mastery over a dataset, acknowledging that neither individual intentions nor current conventions circumscribe how data might be understood or received, and that understanding and reception are both contested and dynamic. It means accepting that correct data can be ethically and morally wrong, and that appeals to correctness do not make data creators less accountable for what their data says.

I can see how the idea of owning one's artistry this way, as a data creator, might seem paralyzing. That's why my students yearn for the world to be otherwise. But although it's comforting to imagine a reality in which data creators follow straightforward and manageable processes to collect facts, that is never what actually happens. And let's make no mistake: misplaced confidence in sufficiency of apparent facts is dangerous. Misplaced confidence in the sufficiency of apparent facts is what enabled the Swedish Institute's data to justify racist policies, such as compulsory sterilization for "medical and social reasons" as Kjellman describes.

So how to move forward, past the raw yearning for facts, to begin the process of owning one's artistry as a data creator, in a very broad sense of that term? Across the adventures in this book, the overlapping paths provide a sense of direction. Over and over, this book has emphasized the human character of data, not only in demonstrating its pervasive ambiguity, but in finding alternate forms of value within that very messiness and uncertainty. In positioning the human character of data as a precious rather than a pernicious quality, this book also argues for according greater dignity to the activities of data creators and showing greater respect for human judgment in the context of data work. Data creation is not mechanical; it requires skill, creativity, and care.

This book also suggests a revised understanding of data constraints: data standards, data collection protocols, data structures, and the like. We need these devices, but not because they enable us to remove the human character from our data. We need these devices because they provide a framework to surface, reflect on, and wrestle with data's human character; they provide a medium through which the process of owning one's artistry can begin. Concurrently, the book illustrates how sustained attention to the structural elements of datasets—how they are put together, from a conceptual perspective—enables the ongoing interrogation of that human character. Such exercises in critical reading are necessary to enable any claim of ownership over one's data artistry, even as, in ways that the book likewise continually underscores, our critical epiphanies are inevitably limited in scope and duration, always subject to reassessment and reversal.

Finally, the adventures in this book demonstrate that the endless process of owning one's data artistry can be invigorating—galvanizing, even. Each ramble over the data landscape, mulling over the here and there of everyday data, can lead to a wider and more wondrous view.

NOTES

INTRODUCTION

1. Feminist standpoint epistemology originated in the philosophy of science. It responds to the empirical insufficiency of an omniscient, putatively objective perspective that reveals itself under scrutiny merely to normalize the particularities of a dominant group. Feminist standpoint epistemology proposes that a knower is always *situated* in relation to an object. Understanding the position of the knower is necessary to understand the knowledge being produced. Reflexivity on one's perspective—to understand its partial, limited nature—is therefore a fundamental component of knowledge production. Likewise, social identities and relations, such as those associated with race, class, gender, and sexual orientation, contribute to an epistemic agent's capacities to know. Such differences are to be valued. In particular, marginalized communities can be better positioned to question assumptions that the dominant group fails to notice. Sandra Harding's (2004) edited volume compiles an array of foundational articles in feminist standpoint epistemology and is a nice introduction.

2. Kristen Hogan's essay about the project she began during our independent study received the 2010 Miriam Braverman Memorial Prize from the Progressive Librarians Guild (Hogan 2010).

3. In addition to Gloria Anzaldúa, other scholars who have been a beacon for me in synthesizing *how* and *what* include Elizabeth Chin (2016), Mary Ebeling (2016), Saidiya Hartman (2007), Donna Haraway (1988), Annemarie Mol (2002), and Anna Lowenhaupt Tsing (2015).

CHAPTER 1: SERENDIPITY

1. I use the term *organizing system* to mean any set of things—physical things, digital things, or data that describe physical or digital things—structured via principles of selection, description, and arrangement. I use *collection* or *dataset* in a similar sense. The principles of selection, description, and arrangement in any organizing system can be hazy and implicit, or they can be systematic and precise. Robert Glushko also uses the term *organizing system* in his textbook *The Discipline of Organizing*. Glushko (2013) emphasizes that an organizing system is intentionally brought together, but

I don't imply that in my usage. Sometimes organizing principles can be perceived without apparent intention behind them.

2. The Museum of Jurassic Technology in Los Angeles is a more established predecessor to the Museum of Natural and Artificial Ephemerata. Weschler (1996) profiles the Museum of Jurassic Technology and its founder, David Wilson.

3. The Armed Conflict Location and Event Data Project (ACLED; https://acleddata.com) has been collecting information about conflict events since 1997. Initially an academic project focused on Africa and parts of Asia, ACLED is now a nonprofit organization with a worldwide scope. ACLED data as well as data collection protocols (documented in a *codebook* according to the terminology of social science research) are freely available on the Web. In 2020, the Johns Hopkins Coronavirus Research Center database (https://coronavirus.jhu.edu) quickly became an authoritative global resource for comprehensive, up-to-date statistics about the COVID-19 pandemic.

4. For an overview of serendipity from the standpoint of information seeking, see Foster and Ellis (2014). Discussions of browsing include, among many others, Bates (1989) and Marchionini (1995). For information encountering, see Erdelez (1997, 2005).

5. For the activities of early modern bibliographers such as Conrad Gesner and Antonio Possevino, see Besterman (1936), Balsamo (1990), and Blair (2010). Library classificationists include Sayers (1915), Richardson (1930), Bliss (1929), and Mills (1960). Vickery (1960a) introduces design principles for faceted indexing vocabularies. All these classificatory devices employ aboutness as the primary organizing principle. For useful contrasts to the centrality of aboutness, see Lee (2012) on bibliographic classification in ancient China, and Littletree, Belarde-Lewis, and Duarte (2020) on indigenous knowledge organization.

6. To declare that "data is not neutral" has become almost passé in these fields. For examples of scholarship that demonstrates this perspective, see Drucker (2013) for digital humanities, Gitelman et al. (2013) for science and technology studies, and Noble (2018) for information studies, among many others.

7. When scholars from other disciplines have found value in information science, the particulars of data cooking are often an interest. For instance, when information historian Laura Skouvig (2020) examines information management practices in the Copenhagen police department in the late absolutist period (1784–1849), she is seeking to understand how certain kinds of activities involving certain kinds of persons become encoded into a centralized information infrastructure—a ledger with various formal and structural conventions.

8. For overviews of infrastructure and infrastructure studies, see, for instance, Star and Ruhleder (1996), Edwards et al. (2009), and Plantin et al. (2018).

9. As an example of a journalistic account, Angwin et al. (2016) demonstrate how judicial risk-assessment algorithms systematically disadvantage Black defendants.

As a more scholarly example, Noble's (2018) *Algorithms of Oppression* details how search results in various information systems manifest racial stereotypes.

10. Just a few examples that describe how data can enact particular perspectives on the world include Caswell (2012) for archival studies, Currie et al. (2016) for critical data studies, Adler (2017) for information studies, and Dourish and Gomez Cruz (2018) for science and technology studies and human-computer interaction.

11. Ruha Benjamin's *Race After Technology: The New Jim Code* synthesizes example after example in which "objective" systems reinforce White supremacy. For instance, regarding predictive policing technologies, Benjamin observes that "If we consider that institutional racism in this country is an ongoing unnatural disaster, then crime prediction algorithms should more accurately be called crime production algorithms" (Benjamin 2019, 56).

12. Two "poetic" examples that exult in the power of data to describe new worlds are King (2008) and Eco (2009). Artists, too, have been poetically inspired by data (Vesna 2007). As the presence of data has multiplied in the world, so too, suggests Heather Houser (2020) has its potential as an aesthetic force. Data does not need to be understood analytically, argues Houser, to be inspirational; indeed, in the context of environmental artists the experience of what she calls *infowhelm* has been generative.

13. In contrast, the predominant goal, when "biased" data is discovered, is to "eliminate the bias"—perhaps by including a wider range of data sources—rather than reimagining the whole enterprise. As Anna Lauren Hoffman (2020) observes, this calming rhetoric of inclusion is dangerous: expanding the data pool does not, by and large, "fix" data regimes that cause violence to marginalized communities. Along similar lines, the authors of the Feminist Data Manifest-No (Cifor et al. 2019) demand that "data can—and should always—resist reduction. Data is a thing, a process, and a relationship we make and put to use. We can make it and use it differently."

14. The consequences of a complacent attitude toward information seeking can be severe, if often unrecognized by the searcher. For example, sociologist Francesca Tripodi (2022) describes how conservative Christians in the United States "fact check" news sources via verbatim search queries pulled from the sources that searchers are attempting to "verify"; the ensuing results, which reflect the terminology of the query, typically confirm partisan assertions. The situation that Tripodi presents is a matter of information practices (that is, human activities with information) rather than poor information outcomes (that is, something fixable by making adjustments to search engines). As Srinivasan, Finn, and Ames (2017) emphasize, technical fixes that adopt "information determinism"—that is, an assumption that the provision of better or different information will result in desired outcomes—typically fail to take social structure and human agency into account.

15. Within the field of human-information interaction and retrieval, efforts to account for and support this labor are most salient in the domain of exploratory search, in which the information seeker's goals are wider, more complex, and less

well formed (White and Roth 2009). Exploratory search can be conceptualized as a form of learning or understanding (Rieh et al. 2016; Bawden and Robinson 2016). Researchers in human-information interaction have developed interface enhancements and tools to support the activities of discovery and synthesis that exploratory search often requires (e.g., Capra et al. 2015; Crescenzi et al. 2021).

CHAPTER 2: OBJECTIVITY

1. The information scientist and philosopher Patrick Wilson's book *Second-Hand Knowledge* is a detailed inquiry into the notion of cognitive authority: how we trust and accept information when most of what we "know" comes from what we are told by others (Wilson 1983).

2. The attraction of quantitative data is one motivation for personal self-tracking, as described in Deborah Lupton's *The Quantified Self* (2016).

3. In her classic study of business communication genres, JoAnne Yates (1989) describes how the use of numerical summaries in the form of charts, graphs, and descriptive statistics facilitated the development of management strategies focused on easily graspable quantitative metrics.

4. 91 seems like a wonderful mark to me, but scales are always interpreted differently. Even scales that are supposed to be standardized are not interpreted in static fashion, and ideas of what values mean can change, either by convention or through institutional intervention. For example, the American Heart Association adjusted its understanding of hypertension in 2017, so that a blood pressure reading of 130/80 was considered to be *stage 1 hypertension* (in other words, high blood pressure), whereas 130/80 had previously been described as *prehypertension* (not yet high blood pressure).

5. In the United States, the use of quantitative metrics to inform predictive models for judicial risk assessment—for instance, which defendants should be charged with what, or which defendants should be granted bail—is one example recently in the public eye (Angwin et al. 2016; Simonite 2019).

6. In the spring of 2020, two closely aligned grading scandals provide additional heft to this suggestion. When the COVID-19 pandemic made it impossible to hold traditional exams for two large systems of educational certification—A-levels in the United Kingdom and the International Baccalaureate (IB) program, which provides a standardized set of achievement metrics for students around the world—the administrators of these systems looked to technical solutions for implementing grades rather than cultural solutions for reimagining grades. Absent the testing components of their scoring systems, provisional scores for British A-levels and for IBs were algorithmically adjusted based on the level of previous alignment between teacher assessment and final scores, for different schools. In a nutshell, students' provisional scores, as assigned by their local teachers, were raised or lowered on the

basis of prior performance at their schools: whether students at their schools in previous years had done worse on the exam than their teacher's assessment would suggest, or better. In practice, this meant that students at lower-performing schools, typically schools with higher proportions of Black and Brown students, had their final scores lowered. Because university admissions in the United Kingdom and other parts of Europe rely primarily or solely on scores, these algorithmic manipulations had substantial effects: when scores were lowered, for instance, students provisionally admitted to universities were denied admission on the basis of their adjusted scores. After vociferous protests, scores in both systems were readjusted, but the effects to university admissions could not be entirely remediated. Although the decision to adjust scores for individual performance on the basis of historical patterns may seem shocking and inscrutable, precisely the same reasoning has been employed in automated criminal risk assessment, as discussed in note 5. But to my mind this situation underscores the brittle arbitrariness of all grading regimes. Why do we continue to accept them? See Adams (2020) and Satariano (2020) about the A-level situation and Broussard (2020) about the IB situation.

Somewhat ironically, the COVID disruptions that motivated algorithmically adjusted scores for university admissions in the United Kingdom made it easier at some universities to experiment with alternate grading possibilities, such as Pass/Fail. Inspired by such emergency policies at my own university in North Carolina, I was finally able to eliminate quantitative scoring for my courses in the spring semester of 2021, when I returned to the United States. Although in hindsight this seems obvious, I was surprised to realize that making this change enabled me to write assignment feedback in a completely different way, more oriented toward encouraging students to think about alternate possibilities ("What would happen if you did it this way instead?") rather than providing evidence for this or that score ("Because you didn't do what I asked, your score is lowered"). Both of my classes that semester were designed for master's students, and, as a group, the graduate students found the lack of grades easy to accept in the midst of a pandemic. Of course, they were all going to get Ps anyway. But a number of undergraduates were enrolled in my classes that semester also, and they tended to feel unmoored without getting a score. They wanted As. This is completely understandable. The anxieties of late-stage capitalism make it seem as if insufficient quantitative evidence will be a detriment in the job market, and this assumption is hard to refute.

7. Precision and recall were first proposed by Cleverdon (1967). Precision and recall remain fundamental assessment metrics for IR systems today, although additional metrics have arisen to supplement them. Many classic papers from information retrieval, including Cleverdon's, are collected in Karen Spärck-Jones and Peter Willet's (1997) volume *Readings in Information Retrieval*. As a complement, Diane Kelly (2009) provides a comprehensive, thoughtful synthesis of user-based evaluation methods for IR systems.

8. The Southern Poverty Law Center (2017) provides a particularly vivid example of relevant but racist search results as contributing to the radicalization of convicted mass shooter Dylann Roof, who killed nine Black people at the Emanuel African Methodist Episcopal (AME) church in Charleston, South Carolina, in 2015. Safiya Noble (2018) discusses the conditions under which such search results are perfectly in line with assessment metrics such as precision and recall.

9. Metaphor as a cognitive (rather than rhetorical) mechanism is canonically described in Lakoff and Johnson (1980).

10. The transfer of genetic material from one cell to another is, for example, a "meaningless" form of information exchange that can be described with MTC (Gleick 2011).

11. To compound matters further, MTC became more popularly known as *information theory*, but Shannon and Weaver use the term *information* in a specific technical sense that, again, has nothing to do with meaning but only with signal transmission (Gleick 2011).

This has caused endless confusion, because *information* is much more commonly defined as that which informs us, or that to which we ascribe meaning. A common formulation describes the difference between data, information, knowledge, and wisdom as an ascension up a pyramid of meaning. *Data* in this formulation is any kind of discontinuity in the world: a hole in the sand, a point of light in a pool of darkness, or the difference between the shape of an *0* and the shape of a *1*: anything that differentiates something from some other thing. *Information* is meaningful data: the green light that shows a surveillance camera is on, or the number that appears when I step on a scale. *Knowledge* is information that has been verified and synthesized (e.g., my knowledge of bread baking combines written information from food science and cookbooks with practical information from the experience of baking myself). *Wisdom* usually involves the application of knowledge according to some accepted set of values, such as to accomplish social goals, to maximize profits, or to maintain happiness. Ergo I might notice a white patch on the loaf in the bread box this morning (data), identify it as most likely mold (information), determine that mold should not be eaten (knowledge), and then debate internally whether it's acceptable to cut away the moldy part or if I should throw out the loaf entirely (wisdom). It is worth noting that although definitions of data, information, knowledge, and wisdom can be articulated in what appear to be eminently sensible and systematic formations (as in, e.g., Floridi 2010), these concepts, rather like relevance, mutate from author to author, so that one person's data is another's information, one person's information is inclusive of nonfalsifiable statements whereas another's is restricted to "facts," and so on.

12. Studies that document the processes by which context leaches out of data are legion and appear across various academic disciplines. A few of my favorites include Latour (1999), Bowker (2000), and Shavit and Griesemer (2011), all three

from science and technology studies; Goodwin (1994) from anthropology; and Atici et al. (2012) from archeology. More recent work in critical data studies, such as Loukissas (2019) and D'Ignazio and Klein (2020), underscores these earlier findings, exhorting data analysts to take account of a dataset's history. Similarly, Bates, Lin and Goodale (2016) have proposed a methodology of "data journeys" for social science researchers to trace the character of data as it travels from system to system.

13. In academia today, there is a good deal of interest in decentering the human as we attempt to understand the complex relationships between entities (things, animals, forces) in the world. Such perspectives encourage us to consider, for instance, appley-ness from the apple's position. To the extent that such exercises might compel us to imagine entities and relations in different ways, I find them valuable and even inspiring (e.g., Hayles 2014). But I hesitate to imagine that such speculations *really* get us closer to what it means to be an apple or anything else. It seems arrogant to suggest that human beings might achieve this kind of understanding, in terms of being able to let an apple speak for itself.

14. The project uses using the Video Game Metadata Schema (VGMS) standard developed by Jin Ha Lee and her colleagues. Lee, Fox, Cho, and Perti (2013) describes the development of VGMS version 2.0, used in the course. Current and previous releases of the VGMS (now on version 4.0) are available via the GAMER research group Web site: https://gamer.ischool.uw.edu/releases/.

15. I have discussed some of these Metadata Architectures projects in Feinberg (2016) and Feinberg (2017a).

16. Lucy Suchman (2002) describes this in relation to document coders (otherwise known as litigation support) in a legal firm. Today, the world of online content moderation is similar (Roberts 2019). As a counterpoint, Marijel Melo and Laura March (2022) find that master's students applying newly learned skills in a library makerspace appreciated being able to "authentically experience" tedious aspects of the making process such as the "frustration of waiting for hours to use a laser cutter and figuring out how to clean 3D printer heads when they become jammed," because they ended up with a project that visitors to a public showcase could visibly enjoy interacting with. Significantly, as Marshall and Melo (2020) explain, the *making* domain is coded as White and male, in addition to technical, and makerspaces have not been perceived as welcoming to diverse audiences. As another counterpoint, Isto Huvila, who has extensively studied the information practices of archeologists, has shown that the "report"—a synthesis document of an archeological project—remains the standard mechanism for communicating archeological information in Sweden, rather than the data summarized in the report (Huvila 2016). Notably, archeologists find a certain satisfaction in creating the report, even though, as Huvila contends in subsequent work, the "authorship" of archeological reports is perhaps better characterized as the practical responsibility of a manager for a project rather than the relationship between a literary author and a text

(Huvila 2019). The production of an archeological dataset does not inspire similar satisfaction.

In a study of Wikipedia edit-a-thons, March and Dasgupta (2020) observe that facilitators of these events focus less on procedure and technique and more on motivation: often, edit-a-thons are organized to address systemic deficiencies in Wikipedia's composition, such as a preponderance of articles about male artists rather than female artists. Even if an edit-a-thon contributor's changes are reversed, the contributor may find satisfaction in being part of a communal act of repair and resistance. A connection to a larger cause—and community—may help ameliorate some of the drudgery associated with data creation.

Conversely, another strategy might focus on surfacing and celebrating unexpressed design ideas, judgments, and techniques. For instance, Andersen and Wakkary (2019) describe a series of Magic Machine design workshops, in which participants respond to a conceptual prompt (such as "draw a sound") and then implement the prompt using basic craft materials (such as "create a machine to express the sound that you drew, using plastic cups, paper plates, cutlery, and string"). Each "machine" is subsequently presented to the group. The Magic Machine workshop demonstrates how apparently trivial work can be treated seriously as deeply revealing and inspirational. The communal aspect, again, seems important.

17. Nathan Ensmenger (2010) tells how computer programming transitioned from low-status work performed by women to high-status work performed by men. Initially, programming was a task akin to the mathematical computation performed by women *computers* as described in, for instance, Shetterly's popular *Hidden Figures* (2016).

CHAPTER 3: EQUIVALENCE

1. In the United States, butter is typically sold in sticks of 4.0 ounces. However, the shape of the sticks differs in the eastern and western parts of the country. In the west, butter sticks are shorter and fatter, whereas in the east, butter sticks are longer and skinnier. Machinery for packaging butter was introduced in different regions at different times, and changes in the machinery led to changes in the shape of butter sticks (Andres 2014). Today, local dairies keep to the established local shapes. Some dairies with national distribution (such as Land O Lakes) produce regional shapes: short sticks in the west and long sticks in the east. Other national distributors (such as Trader Joe's) produce only one shape (Trader Joe's originated in California, and its butter is short and fat).

2. Throughout this book, I use the terms *class* and *category* interchangeably.

3. Philosophers use the vocabulary of *types* and *tokens* to talk about such distinctions, as summarized in Wetzel (2018). Philosophical discussions tend to focus on an abstract understanding of types and tokens (e.g., are types universals?). Here, I am interested in how we imply these concepts in everyday speech and actions.

4. For all you ever wanted to know about Girl Scout cookies, see Rao (2017) and *Los Angeles Times* (2015).

5. Deborah Turner (2010) describes the phenomenon of an *oral document*, which is not a recorded audio file but a time-bound episode of oral discourse with certain characteristics that stabilize it as a form of document.

6. The idea of the *work* is especially important in textual studies, a branch of literary scholarship that traces versions of texts and provides scholarly rationale for the creation of new editions (Tanselle 1990).

7. FRBR doesn't provide an example that involves the staged production of a play, but it does provide examples in which musical scores and recorded performances of those scores are expressions of the same work. It also provides examples in which filmed versions of plays are new works.

8. Ben Brantley (2017) reviewed such a production (*Hamlet* set in imperial Persia) in the *New York Times*.

9. Sometimes people decide that FRBR fails for a particular kind of information object, such as video games (McDonough et al. 2010) and sometimes people decide that FRBR fails because it has not modeled the idea of the work properly (Renear and Dubin 2007).

10. The tradition of authorial intention as the primary organizing principle for a *work* was most recently championed by G. Thomas Tanselle (1990).

11. Examples of textual scholarship that emphasize the history of versions as contributing to the character of a work include McKenzie (1999), McGann (1983), and Eggert (1999, 2019).

12. Find Austlit at https://www.austlit.edu.au/.

13. The EpiPen price controversy was widely reported in American news media. Pollack (2016) provides an overview of the pricing debate, as well as an account of the generic product's release. Duhigg (2017) describes how, a year later, he was charged $600 for a nongeneric EpiPen two-pack when refilling a prescription for his son. Duhigg then explicitly requested the generic version and was told that he would need to wait 90 minutes while his doctor was contacted for an authorization. Ultimately Duhigg paid $370 for the generic EpiPen. In 2018, another pharmaceutical company, Teva, was approved to sell its own generic version of the EpiPen (Kaplan 2018). Other epinephrine autoinjectors exist (such as Adrenaclick); these are equivalent in the EpiPen in terms of outcome but may use a different delivery system and have different usage instructions (Skinner 2016).

14. In the practical literature of cultural heritage data, Cataloging Cultural Objects (CCO), a content standard for museum description distributed by the Visual Resources Association (VRA), constitutes a notable exception to this tendency (Baca et al. 2006). Although items in museum collections tend to be unique physical objects, they are sometimes parts of collective works (e.g., the components of a

tea set) or representations of other works (e.g., an architectural model of a building or a plaster copy of a Greek statue).

15. Marc Ereshefsky (2000) summarizes various species concepts, along with competing approaches to establishing taxonomic evidence, as part of a philosophical argument that advocates for species pluralism as enabling better science. (Chapter 5 also discusses biological taxonomy and approaches to it.)

16. Although information scientists such as Buckland have examined the nature of documents, this work has remained somewhat separate from the more general, and particularly the more practice-oriented, literature of knowledge organization. Document studies seem to be characterized as abstract and theoretical, even as they are inevitably quite empirical. An early investigation of a specific kind of document is Marcia Bates's (1986) article "What is a reference book?" Buckland's (1991) "Information as thing" is more general but equally concrete. Ron Day's books *Indexing It All: The Subject in the Age of Documentation, Information, and Data* (2015) and *Documentarity: Evidence, Ontology, and Inscription* (2019) provide more encompassing, speculative treatments. In contrast, Ciaran Trace (2017) and Tim Gorichanaz and Kiersten Latham (2016) promote an experiential understanding of documents. Although situated in different literatures, these phenomenologically inspired perspectives have conceptual similarities to David Levy's sociotechnically oriented *Scrolling Forward* (Levy 2016).

17. The three objectives for a library catalog, as first set down by Charles Cutter in 1876, also include "showing what the library has" for a given author, subject, or genre (Cutter 1904.)

18. The current version of cataloging rules used by most Anglo-American libraries is called Resource Description and Access (RDA); it is maintained by an international committee called the RDA Steering Committee.

19. For a primer on cataloging, see Chan and Salaba (2016).

20. For instance, cards for the textbook *Botany: An Introduction to Plant Science*, 7th edition, by James D. Mauseth, published 2009 by Jones and Bartlett, would appear in a card catalog under *B* for the title (*Botany*) and under *B* for the subject (Botany) and under *M* for the author (Mauseth) but not under 2009 or *J*. A particular library might have had separate cabinets for author, title, and subject searching (that is, all the cards by author in one set of cabinets, all the cards by title in a separate set of cabinets, all the cards by subject in a third set of cabinets), or the cards might have been interfiled (by author, by title, and by subject all together).

21. Library catalogs began the process of digitization early; the MARC format for cataloging data was developed in the 1960s. In the first online catalog systems, search was still implemented via the access points—that is, instead of a keyword search that would look for the words *William Shakespeare* anywhere that it might appear in an entry, one would select an author, title, or subject search first. In an

author search, for instance, one would first locate the correct *William Shakespeare* via a search of the author file and then retrieve all the entries associated with William Shakespeare (1564–1616) as an author. (This process is exactly like searching a physical card catalog that has been separated into different sections for author, title, and subject arrangements.) Once one is acclimated to it, this two-step process is more powerful and precise than keyword retrieval, but it feels cumbersome to the uninitiated. Catalog interfaces today prioritize simple keyword searching, which just locates a string anywhere in the entry, rendering the access point terminology meaningless.

22. Although bibliographic classifications were focused on physical arrangement, the intellectual order provided by such a system of relationships can also structure digital environments, as well as more abstract forms of knowledge representation (e.g., the *ontologies* of computer science). (Chapter 5 addresses the utility of taxonomic structure more generally.)

23. Three complementary publications produced by members of the British Classification Research Group (CRG) are prototypical examples of guidance for such specialized classifications (or indexing languages): *Classification and Indexing in Science*, *Classification and Indexing in the Social Sciences*, and *Classification and Indexing in the Humanities* (Vickery 1975; Foskett 1974; Langridge 1976). Classification nerds like me revere the CRG. For the nonspecialist, it can be challenging to navigate the technical particulars that riddle the CRG's work—discussion of notations and filing order can seem mystifying today—but the CRG's efforts to develop coherent and elegant, as well as pragmatic, classificatory devices remain exemplary.

24. Put another way: as devices of data infrastructure, bibliographic classification schemes (used to arrange items on library shelves), indexing languages (used to describe research articles in databases like PubMed), and subject headings (used to describe items in library catalogs) are all controlled vocabularies used to generate aboutness data. These fundamental similarities have been occluded by divergent institutional practices.

25. In contrast, Bernhard Rieder's *Engines of Order* characterizes the structural difference between precoordinate, enumerative classification schemes and postcoordinate indexing languages as a definitive conceptual break (Rieder 2020). In a *precoordinate* scheme, classes like "Tensile properties of carbon-steel alloys" are instantiated as a compound, and all classes must be explicitly defined by the classificationist (the scheme designer). In a *postcoordinate* system, each component of a complex subject— tensile properties, carbon-steel, alloys—is defined separately and can be combined at will by the classifier (the person using the scheme to describe a document). From my perspective, Rieder underplays the conceptual heritage that postcoordinate indexing shares with faceted (or analytico-synthetic) approaches to bibliographic classification, most notably in the work of S. R. Ranganathan (Classification Research Group 1955).

26. Lilly Irani (2019) describes how this valorization of design thinking operates to maintain Western dominance in globalized labor markets.

27. The Dublin Core is a standard for describing informational things (*resources*, in Dublin Core terminology). The Dublin Core is maintained by the Dublin Core Metadata Initiative (DCMI)—which is just a group of interested people, similar to the World Wide Web Consortium (W3C). (See the DCMI Web site at https://www.dublincore.org/.) The DCMI describes and promulgates a variety of best practices for data collection. Dublin Core, which has been codified as both an American (ANSI) and international (ISO) standard, has achieved widespread usage as a lingua franca for basic data about information resources. Sometimes, data about information resources is called *metadata;* thus the DCMI, and not the DCDI. (See chapter 5 for more about metadata in comparison with data.)

28. Indeed, as Andrea Thomer and Karen Wickett (2020) productively observe, the line between *dataset* and *database*, if there is one, is equally porous in the context of scientific data collection. Does a dataset describe one thing, while a database relates descriptions of multiple things? In practice people implement databases (and datasets) with a wide range of goals, in a wide range of structures and formats. Importantly, as Thomer and Wickett emphasize, ambiguity in these situations is not the result of sloppiness. The case studies that they analyze include compendia of *decades* of research—the lifework of scholars. Similarly, Montoya and Morrison (2019) contribute an equally thorough account of decades' worth of archeological data from the Angel Mounds site in Indiana, describing how subsequent teams of data collectors employed a consistent identification mechanism for field specimens, and yet the aggregated "database" of site data consists of a series of linked datasets that overlap in different ways.

29. As one example, Couldry and Mejias's (2019) book on data colonialism warns that our "inner thoughts" are "made ready for appropriation as data" in a manner similar to the exploitation of natural resources by colonial powers, such that "human life itself" is colonized via the collection and aggregation of "personal" data. Which is not to say that we shouldn't be very worried about the collection and aggregation of digital traces but only that the equation of personal data with inner thoughts, life itself, and, indeed, "me" might be further interrogated.

30. Johanna Drucker (2006) articulates the default orientation of literary scholars thus: "Most are closet transcendentalists, harboring a not-so-secret belief that after all it is the 'sense' that 'really matters.'" Digitization has encouraged a reassessment regarding the material basis of that "sense," in terms of digital versions of print works (e.g., Mak 2011), digital works (e.g., Drucker 2013; Kirschenbaum 2008; McGann 2001), and nontext modalities (e.g., Clement 2016). Some of this scholarship reimagines the locus of the work quite significantly; Alan Liu, for instance, proposes that all literary works—digital and not—are networks embedded in other networks (Liu 2018). As a complementary perspective, Nanna Bonde Thylstrup

(2018) proposes that mass digitization projects in the realm of cultural heritage (that is, libraries, archives, and museums) subsume individual works into infrastructural assemblages whose role, increasingly, is to provide data for subsequent processing and transformation.

31. In human-computer interaction, discussions of physical vs. digital things include Whittaker and Hirshberg (2001), Kirk and Sellen (2010), Dourish and Mazmanian (2013), and Rosner (2012). Work oriented more toward digital materiality includes Dourish (2014) and Odom, Zimmerman, and Forlizzi (2014). (I too have published in these areas. Feinberg [2013] and Feinberg, Broussard, and Whitworth [2016] concern physical and digital things, and Feinberg [2017b] and Feinberg, Carter, Bullard, and Gursoy [2017] focus more on digital materiality exclusively.)

32. Sociological perspectives emanate from actor-network theory, as articulated by Bruno Latour and John Law, in which nonhuman actants such as a door closer, navigational instruments, or even the wind and weather participate equally in the unfolding of historical events and social practices (Latour 1988; Law 1984, 1992). In the philosophy of science, Donna Haraway and Karen Barad likewise emphasize the entanglement (to use Barad's term) of human and nonhuman, so that it is impossible to understand each independently (Haraway 1988; Barad 2007). Other philosophical accounts include postphenomenology (e.g., Verbeek 2005) and object-oriented ontology (e.g., Bogost 2012). These latter directions have spurred some empirical work in human-computer interaction in which, for instance, researchers might try to understand the world from the perspective of a scooter on the streets of Taiwan (Chang et al. 2017).

33. As put by sociologists Law and Lien (2013), "Ordering becomes a relational and performative effect of practices, and since the latter vary, this also means that ordering varies too." In Law and Lien's study, farmed Atlantic salmon at a Norwegian fishery are different entities from salmon in their wild habitat (or salmon in a biological taxonomy). In a few other examples of such studies, atherosclerosis in the clinic and atherosclerosis in the pathology lab are different conditions (Mol 2002), and Amazonian forests are thinking beings (Kohn 2013). Such work asserts an ontological perspective rather than a constructivist one; that is, it is not just that Norwegian aquaculturists think about salmon in a certain way but that the salmon that they farm are empirically distinct ontological units.

34. For discussions of warrant, see Beghtol (1986, 2002) and Bullard (2017).

CHAPTER 4: INTEROPERABILITY

1. The way that recipes tend to be written now—with precisely measured lists of ingredients, numbered steps, and fully described outcomes—makes it seem as cooking is about following instructions precisely. (In contrast, older cookbooks include less detail and seem incomplete to the modern reader.) But of course this

is bosh. Even recipes that provide comprehensive rationale for their method (e.g., *Cook's Illustrated* or the *Food Lab*) can't anticipate what will happen with your equipment and materials. In response to these contradictions, recipe readers operate within a broad spectrum of literalness, in ways that seem normal and reasonable to one person but crazily idiosyncratic to others. One only needs to look at online comments for any recipe site. Invariably, someone posts that a recipe is fantastic, adding that they changed six of ten ingredients and opted to steam instead of fry. "But that's a completely different dish," another reader will respond, and then a third person will say, "Good grief, it's only pasta," or whatever it is. Meanwhile, another poster will complain that they followed the steps to the letter, but the Brussels sprouts never browned, and what did they do wrong? Even as we believe that we are closely following a documented standard, we are seldom doing so to the degree that we think we are.

2. As a White, middle-class person, my childhood "small world" was quite permeable, as my story demonstrates; easy to breach, without risk to myself or others. In contrast, the notion of "small world" developed by information scientist Elfreda Chatman involves much stronger barriers (Chatman 1991, 1996, 1999). Chatman studied information poverty in communities such as university custodians, retirees, and prisoners. Her research participants were primarily working class, poor, BIPOC women, often mothers. For Chatman's participants, going outside of their "small worlds" could be dangerously disruptive, and they avoided it, not wanting to jeopardize the stability of their current situations. Accordingly, the "insider's world view" that Chatman describes is characterized by predictable routines, "greater reliance on self and a general distrust of outsiders" (Chatman 1991, 445). More recently, Amelia Gibson and John Martin (2019) extend Chatman's ideas about information poverty by placing more analytic focus on people's self-protective information strategies, rather than the conditions that provoke these responses. Gibson and Martin propose that a shift to *information marginalization*, rather than information poverty, "refocuses blame away from individuals experiencing marginalization, and toward the contextual conditions that create information poverty" (Gibson and Martin 2019, 485).

3. The Thai Delicious committee seems to have transitioned into a company that provides standard recipes and sells standardized products (as described on its Web site, https://www.thaideliciousfood.com/en/).

4. It might initially seem as if a concept as starkly dramatic as *terrorism* might be less uncertain than a concept like *French bread* or *Thai food*, but this is not the case, as *New York Times* reporter Margot Sanger-Katz discovered in 2016. Data projects that track worldwide violence may classify the same event differently, depending on both the concept definition that they use and on the application of that concept to actual events. When databases disagree on whether certain events constitute terrorism, choosing one database over another can lead to different conclusions as to whether terrorism is increasing or not, depending on the geographic area we are

trying to understand. (The incidence of terrorism in most Western countries is very small, and so even a few disagreements can affect how we interpret the rate.)

5. For instance, Taub and Fisher (2018) summarize Facebook's connection to eruptions of violence in countries such as Sri Lanka and Myanmar.

6. Dangerous posts often involve the propagation of falsehoods—either misinformation (unintended errors) or disinformation (deliberate deceptions). These falsehoods are often loosely referred to as *fake news*. One might imagine that less judgment is required to identify and remove untruths. But ecosystems of misinformation and disinformation—fake news—are intricate and complex. Jack Andersen and Sille Obelitz Søe (2020) propose that fake news must be understood as a communicative action like any other utterance and, accordingly, that the identification and elimination of fake news cannot be algorithmically determined. Amelia Acker and Joan Donovan (2019) call attention to the sophisticated set of techniques by which communicators of disinformation disseminate their messages across platforms. Acker and Donovan describe these techniques as *data craft*—which I read as an analogy to spycraft. Data craft adepts require detective work to be unmasked, and putting the clues together requires judgment and skill. In such conditions, Carter, Acker, and Sholler (2021) suggest, scholars might consider looking to investigative journalism to inform research methodology.

7. S.R. Ranganathan (1959) identified "flair" as a necessary component of a skilled classifier's practice. Ranganathan struggled to define and systematize flair, but it's clear that it involves the skilled application of informed judgment, as opposed to whim or caprice. Flair is, in other words, not a flouting of rules and systems but a reasoned assessment of when and how to apply them. Similarly, Allyson Carlyle (2015) describes how library catalogers are constantly challenged to interpret and extend their rules in new ways, even as a naive outsider might believe that cataloging is merely about memorizing and applying a fearsomely comprehensive set of detailed prescriptions. Both Carlyle and Ranganathan address the information professional—someone with particular expertise who has been educated in a discipline, who is a member of a professional community, with associated best practices, and who has other related qualifications. But as Jaime Snyder's (2017) wonderful study of "vernacular visualization" work in a citizen science organization demonstrates, nonprofessionals also accumulate competencies that enable the development of flair, even as the form that flair might take, and the goals to which that flair might be put, differ from the judgments that professionals might make.

8. Vexingly, it seems impossible to know when and how judgment might need to be applied to ensure that technical devices are implemented appropriately, in human terms. As I was finishing my first draft of this chapter in January 2020, I came across two news items: the first, that a machine learning system had performed better than humans at diagnosing breast cancer from mammograms, and the second, that robots in Japanese factories are terrible at removing eyes from

potatoes and peeling squashes (Walsh 2020; Rich 2020). It turns out that human judgment is really necessary for preparing potatoes and squashes, and less so for routine identification of cancerous tumors. Both of these results were somewhat surprising to the system developers.

9. On the challenges associated with data reuse, see, for instance, Faniel and Zimmerman (2011); Tenopir, Allard, and Douglas (2011); and Wallis, Rolando, and Borgman (2013).

10. Illegal districts might *pack* Black voters into as few districts as possible or *crack* the power of Black voters by distributing majority-Black areas over multiple districts. In 2017, for instance, district maps in North Carolina were invalidated by the US Supreme Court for exactly these reasons (Liptak 2017).

11. Jeff and Shaowen Bardzell have systematically articulated a role for criticism in human-computer interaction, arguing for criticism to be respected as a rigorous enterprise and recognized as a vital component of the discipline (Bardzell 2011; Bardzell and Bardzell 2013; Bardzell and Bardzell 2015).

12. This understanding of criticism as a form of nurturing bears alignments to María Puig de la Bellacasa's proposal for *matters of care* as an alternative to Bruno Latour's *matters of concern* (Bellacasa 2017). Care, in this formulation, continues acts of noticing into acts of support or maintenance, in the service of communal flourishing. We can think of criticism, in this manner, as recognizing the value in something so then to strengthen it. Most academics can relate to this. There is nothing more dispiriting than presenting a paper that you've slaved over at a conference and then getting no questions at all. It's the pits! In contrast, a thoughtful question can help you think about your own work differently, to make progress on it. It shows care.

CHAPTER 5: TAXONOMY

1. This snippet of a controlled vocabulary is taken from the subject categories associated with the Warburg Institute's Iconographic Database, available at https://iconographic.warburg.sas.ac.uk/.

2. Barbara Kwasnik (1999) describes the power hierarchy relationship as a *tree*, using the example of a chain of command in the military.

3. When literary critics analyze hierarchical structure, for instance, their arguments may unfold along similar lines to those of my students. Caroline Levine, in her erudite and eclectic reconsideration of form in literature, takes the power hierarchy—which for her is just *hierarchy*—as a fundamental form, along with wholes, rhythm, and networks (Levine 2015). Levine's hierarchy often takes the shape of a binary division in which one component is dominant over the other (such as a gender binary) but may also describe other structures in which power is

distributed unequally (such as a corporate organizational structure in which power is directed downward in successive layers of management). When Levine looks at hierarchy in Sophocles's *Antigone*, she examines the conflicts that play out between intersecting "hierarchical binaries" that recur in the play: male/female, gods/humans, obedience/disobedience, public/private, friend/enemy, kings/subjects. These forms are hierarchical in the power sense but not in the taxonomic sense.

4. Although the genus:species relationship (or *is a* relationship) is the most common hierarchical relationship, there are others also. One is part:whole (for example, from bicycle to wheels and frame, or from plant to roots, stem, and leaves) and another is kind:instance (so, from oceans to Atlantic Ocean or from cities to Istanbul). But the genus:species relationship is the most important, and that's why I focus on it here.

5. Organizing principles must be constant between each parent class and its children, but the principles can vary across the taxonomy. The principles that relate Air to its subclasses and Carbon Dioxide to its subclasses can be different. Here, leavening agents that introduce Carbon Dioxide into a dough are either Chemical or Biological in nature. The organizing principle that links Carbon Dioxide to its subclasses has to do with whether the agent is organic (Biological) or inorganic (Chemical)

6. It may seem wrong to imagine data that consists of a bunch of mixed-up bodily measurements thrown together without context. How could we possibly know whether that number refers to the circumference of a head or thigh, or whether a number that seems like it must be an inseam is actually the total height of a child, and so forth? And yet in other cases we mix together these kinds of bits without comment, as when a set of keywords might refer to anything under the sun. On an individual basis, keywords might be both accurate and insightful, just like the grab bag of measurements might be. But in aggregate, keywords can be misleading, because each thing is described with different kinds of data. This is not to say that keywords are useless, but merely that it can be easy to blithely ignore the indeterminacy of keywords, even as we might find that same ambiguity troubling in other forms of data.

7. Because I was trying to keep the diagram from getting unbearably cumbersome, most of the subclasses in my reworked gestures taxonomy are not jointly exhaustive. It would be quite an enterprise to describe all types of purposive actions that might be referred to in a gesture.

8. Faceted classification, as initially proposed by S. R. Ranganathan, is an attempt to deal with the problem of multiple principles of division (as described previously) in the context of library classification, in which physical books are arranged in linear order according to their subject matter (Ranganathan 1957; Ranganathan 1959). Occasionally, the notion of faceted classification is held up as a (less evil) alternative to (evil) hierarchical classification. But that is a false comparison. A faceted classification is perfectly compatible with hierarchy. A faceted classification,

which is invariably *synthetic*, is oppositional to *enumerative* classification. (This distinction is explained in the reflection essay for chapter 3.)

9. If you're wondering why the labels in my revamped gestures taxonomy are so long—why I've used "(Gestures that use the) head" as opposed to just "Head"—it is just to be explicit that all the data values are types of gestures. If we were using this taxonomy to generate data about a painting, we are saying that the gesture depicted in the painting takes place on a cloud.

10. The tenor of my argument here bears some similarities to Louise Amoore's (2020) *Cloud Ethics*, although her focus is on algorithms and mine is on the data that algorithms are trained on and operate upon.

11. To increase confusion, once a schema is encoded in a particular way, people are apt to refer to the logical structure of a schema using the terminology of the encoding, which can make it difficult to distinguish these two layers.

12. My own use of the word *taxonomy* has been strategic. Years ago I described this kind of structure as a *classification* when I assigned design projects to my students. But I found that students dismissed their skills when I used this term. They didn't feel qualified to apply for taxonomist jobs at media and technology companies with such competencies. So I changed the name of my project from *classification project* to *taxonomy project*. Voilà! The students now believed they could work at Amazon as taxonomists. Similarly, it has been observed that *ontology* is "classification for men," because *ontology* sounds technical in a sexy way, while *classification* sounds technical in frumpy way. *Ontology* is a word for (male) software architects who get stock options. *Classification* is a word for (female) librarians who live in studio apartments full of cats. Like teaching and nursing, the social status of librarianship has been diminished due to its association with women, to the extent that the library's (sophisticated) technologies are not recognized as proper technologies. For evidence of the scorn accorded to librarianship as a profession, one need look no farther than the classic film *It's a Wonderful Life*, in which the absolute worst permutation of the Pottersville alternate universe is that Mary, wife to George Bailey in the real world, has become *a librarian*. "Noooo!!!" shrieks George, overcome with the horror.

13. What is *metadata?* Sometimes, this term is used to encompass all the conceptual infrastructure of data creation: a schema and its elements, a controlled vocabulary and its values. Other times, this term is used to describe a subset of data—data that contextualizes some other data. If a book is data, the author and title of the book are metadata. If an e-mail message is data, the sender, recipient, subject line, and all the technical information about the transmission are metadata. Both of these usages are confusing. Lots of things can have titles. I just added an event to my online calendar and gave it the title "Zoom with Deb." Is that title data or metadata? There is no additional content to the event in the calendar except for the information about it. So it's data then? So titles can sometimes be data and sometimes be metadata? Yes. Basically, anything can be either data or metadata depending on one's state of

mind at the time. Personally, I avoid the term *metadata* whenever possible. I'd just as soon it didn't exist. (For comprehensive and thoughtful accounts of metadata, see Mayernik 2019 and Mayernik 2020.)

14. I'm writing this during the COVID-19 pandemic, at a time when the tally of a particular cause of death is very important to many people around the world. The ICD is an integral component of such counts.

15. Philosopher Sarah-Jane Leslie (2013) discusses the persistence of ideas of essences and natural kinds in philosophy (via the lingering influence of Saul Kripke and Hilary Putnam) despite its rejection by biological taxonomists and philosophers of science.

16. The indeterminability of species has begun to seep into the general consciousness as well; as one example, a 2017 feature in the *New York Times Magazine* discussed relative closeness of early hominid species under the title "Neanderthals were people, too" (Moallem 2017).

17. In speaking of scientific taxonomy, for instance, John Dupré (2006) observed that "classifications are good or bad for particular purposes, and different purposes will motivate different classifications." The implication is that there is no real distinction between scientific and nonscientific classification and that accuracy, in terms of alignment with observable states of the world, does not entail comprehensiveness, universality, objectivity, stability, and other qualities that we might naively associate with scientific truth as obtained via scientific methods. In information science, although some have attempted to maintain a distinction between scientific classificatory processes and informal, everyday classificatory processes (e.g., Jacob 1991), such defenses have not held up well under scrutiny (e.g., Mai 2011).

18. Henry Bliss, for example, took scientific consensus as the primary design rationale (or *warrant*) for his bibliographic classification, initially developed for the City College of New York. Although Bliss's classification was not widely implemented, he was influential as a classification theorist (Bliss 1929).

19. Because bibliographic classification schemes were employed to arrange library collections, for instance, it didn't make pragmatic sense to create classes for knowledge that was not actually the subject matter of books—e.g., atoms might be the building blocks of matter, but if no one writes books about atoms per se (as opposed to writing about the fundamentals of chemistry or physics), then this *literary warrant* would indicate that atoms should not be included as classes in a bibliographic classification (Hulme 1911).

20. Ranganathan (1957, 1959), trained as a mathematician, was particularly notable in using the language of logical proof for his classificatory recommendations (e.g., "the canon of decreasing extension" or "the principle of increasing concreteness").

21. There are also, of course, historical examples of design experiments that stretch or challenge the idea of taxonomic structure as we understand it today.

When such experiments are not taken up, they tend to fall into obscurity. As Post, Golden, and Shaw (2018) observe in the context of XLink, an extension of HTML from the 1990s, and netomat, an artist-designed browser that made use of XLink, a look back at such experiments can prompt reconsideration of current design orthodoxies. In the taxonomic space, for instance, Barbara Kyle's innovative, discipline-independent approach to concept formation in a faceted taxonomy of social science challenged received notions of fundamental elements and their warrant (Kyle 1958).

22. Redström (2008) discusses the continuation of design through use.

23. Julia Bullard (2018) describes the system of *curated folksonomy* implemented by a popular fan-fiction archive, in which volunteer "tag wranglers" work to identify preferred terms and relate apparent synonyms for the tags that authors freely submit. Behind the scenes, wranglers argue intensely about the concepts at issue in their respective fandoms. Bullard's study emphasizes the scale of the work involved (there are hundreds of tag wranglers) and the seriousness, dedication, and expertise that the wranglers bring to bear in shaping submitted tags into a usable and coherent system.

24. The taxonomy, which emphasizes relationships between classes, or sets, is one kind of structure. The *graph* is another, looser kind of structure, which involves links from one object to another (the Web, with its system of undirected links, is a kind of graph structure). Relationships in graphs are undifferentiated; A and B are connected by a link, but the nature of that link is unknown. (In contrast, in a taxonomy, B is a kind of A.) Scholars such as Tara McPherson have argued that the looseness of graphs can enable radically emancipatory knowledge structures. McPherson developed the digital publishing system Scalar to explore these ideas (McPherson 2018). Scalar provides mechanisms to annotate and relate digital media completely through non-hierarchical associative relationships—an elaborate system of tagging. A team of students and I conducted some experiments with Scalar, in which we found its operation similar to a jumble drawer; freeing in some ways, but conducive more to obscurity than transparency, and almost impossible to maintain, particularly in collaborative environments (Feinberg, Bullard, Carter, and Gursoy 2017).

25. Jo Freeman (1975/2013) observed that the informal structures that arise to fill the vacuum in purposefully "structureless" social groups invariably concentrate power in the hands of a privileged few, even as the motivation toward structurelessness is grounded in emancipatory ideologies. Similarly, sociologist Jeremy Gilbert (2014) found that radical politics are most ably (if tediously) furthered through the structured collaboration of (ongoing!) meetings with attention paid to the details of process.

26. In the popular imagination, all cultural heritage institutions—of which the most prominent are libraries, archives, and museums—have similar aims, all centered

around public access to different kinds of information. In fact, although these institutions might have been more closely aligned toward the beginning of their modern formation, they quickly separated, with each developing its own set of goals, with correspondingly distinct modes of information provision and management (Given and McTavish 2010). Of the three, only libraries emphasized user needs. In contrast, archivists focused on maintaining the integrity of the documents as they received them; to describe or arrange archival materials according to the wishes of potential users was traditionally seen as corrupting their evidentiary value (see, for instance, McNeil 1994; Cook 1997; Gilliland 2000). Perhaps because of these traditions, archival scholarship in the past 20 years has been deeply self-reflective and critical of the archivist's power to define historical narratives (e.g., Duff and Harris 2002). More recently, archival scholars have taken a more activist turn, advocating for participatory, community-based archives as a mechanism to facilitate equity and inclusion (e.g., Caswell 2014; Cifor 2016; Sutherland 2017; Caswell, Punzalan, and Sangwand 2017).

CHAPTER 6: LABELS

1. As a reader of Slate's Dear Prudence column, I was initially surprised to see that "name stealing" was a perennial concern. The latest letter on this topic appeared just as I was finalizing the manuscript, with the headline "Dear Prudence, My Cousin Turned My Family Against Me Over a Baby Name" (Desmond-Harris 2021).

2. Information systems can propagate this kind of violence when they make it difficult to change one's name or eliminate references to a former name. As described by Haimson and Hoffman (2016), Facebook's longstanding policy of demanding "real names" and "authentic identities" resulted in the deactivation of accounts created by trans, binary, and Indigenous users, as well as drag queens and sexual abuse survivors, all of whom employed names different from those on their birth certificates. (Facebook altered its policies in 2018 to address some of these concerns.) Hoffman, discussing the case of trans and gender-nonconforming people, characterizes such policies as *data violence* (Hoffman 2017). Keyes (2020) relates the experience of being deadnamed by an information system in the context of an in-person retail transaction and the accompanying fear that they would be outed as trans to their companion. (The term *deadname* refers to the act of calling a trans or nonbinary person by the name they were given at birth rather than the name that they currently use.)

3. For news reports on Greece and Macedonia, see, for instance, Lewis (1993), Smith (2014); Kitsantonis (2018), Bildt (2018), Santora (2018), and Smith (2019).

4. Of course, although shared names are manageable in everyday interactions, they can cause difficulties when precise identification is required. But even uncommon names are typically insufficient to use as unique identifiers, even in relatively small datasets. The ability to search for one's name on the Internet has made this

apparent. (In Cathy Marshall and Siân Lindley's (2015) study of people searching for themselves on the Web, participants were quite aware of their Internet doppelgangers. There are, for instance, various Melanie Feinbergs around!) But shared names and errant identification based on misapplication of shared names are only more familiar in the Internet era rather than more widespread. Describing her multiyear compilation of documentary evidence related to the biography of Joan Vollmer, most well known as the second wife of writer William S. Burroughs (and shot to death by him in Mexico), Cathy Marshall describes various situations in which unclear or mistaken names enable incorrect information to propagate, sometimes because they were recorded in traditionally published sources. Because social networks embed names within networks of connection, Marshall posits that more precise identification might be easier with social media data than with older ephemeral sources—if social media is archived and made available for research, that is (Marshall 2018).

5. This legal precedent was established in Hamilton v. Alabama. 376 U.S. 650 (1964).

6. Literally translated, the word for *pink* in Danish is *light red*: *lyserød*. So this isn't such a weird fantasy!

7. The actual date of the change is not certain. I own the fourteenth edition of *The Chicago Manual of Style*. I no longer remember exactly when I bought it, but it was probably 1995—that's when I became a development editor and began acquiring a number of reference works. In my copy, *Oriental* does not appear in the examples of "nationalities, tribes, and other groups of people." *Asian* does. But when I looked at digitized copies of the *Chicago Manual* from 1993 and 1994 via the HathiTrust, *Oriental* appears instead (along with a number of other examples that were removed in my version, including *red man* as an example of a term that should be lowercased). My version has the same ISBN as the earlier ones, so mine must have been considered a mere reprint. The list of examples in the earlier versions of the fourteenth edition is very close to the list that appeared in the thirteenth edition from 1982. I also checked a few versions of the Associated Press Stylebook. In the 1987 edition, *Oriental* is included without comment as a term to describe people, but in the 1994 edition, the entry for *Oriental* has been updated with an instruction to prefer *Asian* instead. (There is also a 1992 edition of the AP Stylebook, but it was not available at my university library.)

8. In 2018, the PBS Web series Origin of Everything produced a seven-minute video that encapsulates this history: *Why Do We Say "Asian American" Instead of "Oriental?"* (PBS Digital Studios 2018). Historian Erika Lee's (2015) book *The Making of Asian America: A History* provides an in-depth treatment.

9. For reports on Trump's use of the term *Chinese virus*, and subsequent violence against Asians in the United States, see, for instance, Rogers, Jakes, and Swanson (2020); Tavernise and Oppel (2020); Stevens (2020); and Yang (2020). Anti-Asian

violence continued to escalate in 2021 with incidents across the country, including the shooting deaths of six Asian spa workers in Atlanta (two additional people were also killed in the shootings). See, for instance, Taylor and Hauser (2021), Hong et al. (2021), and Fuller (2021).

10. *Indexing* as a domain of work was initially focused on the description of self-contained units within serial publications (e.g., to describe individual articles within academic journals). In libraries, catalogers had determined to describe only entire works and not their components (e.g., a book and not the chapters within a book, or a volume of a serial publication and not the issues or articles within that volume). Separate indexing and abstracting services then arose to create this data. Initially, these services produced print *indexes* for the literature produced in a given subject area during a specific period. (Perhaps, if you are old like me, you remember using the Reader's Guide to Periodical Literature in elementary school? Each yearly edition grouped articles from general-interest magazines under subject terms. To do one's school report on, say, air pollution, one would look up the term and find the articles from that year. The research databases of today are direct translations of these print indexes into digital form.)

Subject cataloging and *indexing* are essentially similar: both activities involve creating aboutness data for information resources. The unit being described differs (a book, as opposed to an article), as do the controlled vocabularies used to generate the aboutness data. However, despite the similarities, catalogers and indexers developed distinct professional identities, along with their own practitioner literature and professional jargon.

11. Two representative examples of manuals for thesaurus construction are Aitchison, Gilchrist, and Bawden (2000) and Broughton (2006).

12. For more about the Simple Knowledge Organization System (SKOS), now a W3C standard, see its Web site at https://www.w3.org/2004/02/skos/.

13. My thinking generally on concepts, labels, terms, and the whole shebang owes much to Jonathan Furner's (2012) article "FRSAD and the Ontology of Subjects of Works," which astounds me with its brilliant clarity every time I read it.

14. For library classificationists writing about the systems of their design, see, for instance Sayers (1915), Bliss (1929), Richardson (1930), and Ranganathan (1959).

15. For classificationists describing indexing languages of their design, see, for instance, Foskett (1974), Vickery (1975), and Langridge (1976).

16. Berman is not, of course, the only person to have taken a critical position on naming systems in libraries. Hope Olson and Rose Schlegl (2011) usefully summarized an array of such critiques in cataloging literature, focusing on the representation of women. Olson's 2002 book, *The Power to Name*, is an extensive discussion on these topics. A recent treatment is Melissa Adler's (2017) book *Cruising the Library*, which focuses on names pertaining to homosexuality. Michael Buckland

(2012) and Emily Drabinski (2013), provide a counterpoint to these critiques, focusing on LCSH / the library catalog as the continued product of many hands, over time, through a continually evolving society. Although the academic and technical literature around such naming issues is quite extensive, it tends to be little known outside the field. But occasionally a particular naming concern will find its way into the public eye. For instance, in 2016, the Republican-majority House of Representatives voted to prohibit the Library of Congress from changing the subject heading for *illegal alien*. Although subject heading revisions are typically routine—LC vocabularies are continuously updated—this one had already been quite public. After an initial request by students and librarians at Dartmouth had failed to gain approval, the American Library Association had issued a resolution supporting the Dartmouth request; LC subsequently worked out a combination of terms to take the place of *illegal alien*. As noted in *Library Journal*, this case was unusual not merely for the Congressional reaction but because it had originated with an actual patron (a Dartmouth student) noticing a problematic name and pursuing the matter (Peet 2016).

CHAPTER 7: LOCALITY

1. Yanni Loukissas's (2019) book *All Data Are Local* has similar aims to those of this chapter. Loukissas's approach and disciplinary perspective are different; he uses visualization to surface data alignments and misalignments in a variety of settings. Similarly, D'Ignazio and Klein's *Data Feminism* (2020) (chapter 7, "Show Your Work") also has similar goals but an alternate approach; they focus on data labor and the circumstances of production. In a way it's striking how little content overlap there is between this book, *All Data Are Local*, and *Data Feminism*, given the commonalities of our positions. The three works complement each other well!

2. In my memories of attending it from 2010 to 2014, the Texas Book Festival, like Louisiana Literature, was a wonderful event. Both of these festivals program diverse authors and topics, and both events have made efforts to facilitate inclusivity. Louisiana Literature costs only the price of a regular museum admission. The Texas Book Festival is free. Nonetheless barriers remain, as my observations reflect.

3. Like all data, race data can be a double-edged sword. Race data can be used to illuminate and address inequities. Documented income disparities between Black and White families, for instance, can promote compensatory action, such as job training and recruitment programs. But race data can also increase inequities. For example, data about the racial makeup of residential neighborhoods enabled the discriminatory real-estate practices known as redlining (see Koopman [2019] for a data-oriented discussion of redlining).

4. The census implementation of race data is based on standard race categories for US government agencies, as last revised in 1997 by the Office of Management and Budget (OMB). These 1997 categories result in significant numbers of people,

particularly those of Latinx and Middle Eastern and North African origins, selecting "Some Other Race."

5. In the 2020 census, the primary questions remained unchanged, but these were supplemented with an option to write in "origins" next to the selected race boxes (e.g., the instructions next to the White box say to "Print, for example, German, Irish, English, Italian, Lebanese, Egyptian, etc.").

6. The vocabulary that the US Census uses to express race has changed often over its history; here, I discuss only the most recent data implementation along with the most recently proposed revision. In addition to changes in content, structure, and expression, the method of data collection has also changed significantly over time. Significantly, the census has only been self-administered since 1970. Previously, data was collected and recorded by census takers. (Later in this chapter, I do consider how the mode of data collection—e.g., self-identification or independent assessment—affects the construct being described.) Margo Anderson (2015) provides a detailed history of the US Census. Kenneth Prewitt, a professor of public affairs and former director of the US Census Bureau, discusses the census's historical treatment of race in depth (Prewitt 2013). Political scientist Melissa Nobles provides a comparative approach, contrasting how the censuses of Brazil and the United States have collected data about race (Nobles 2001). Literary critic Mike Soto (2016) contributes an account of the Harlem Renaissance informed by a reading of concurrent racial categories as implemented in the US Census: in 1910 and 1920, census takers recorded separate data values for Black and Mulatto (that is, mixed-race) people, whereas in 1930 and 1940, census takers were directed to employ a single data value, Negro.

7. Such practices became much less common following World War II, but they did not disappear entirely. For instance, in the Soviet Union, where religion was not recognized by the state, Jews were identified on passports as an ethnic minority—in other words, as not Russian. (Russia discontinued the identification of ethnic minorities on passports in 1997.)

8. Another way of saying this is that we need to understand how "intuitive" data values work, as well as "nonintuitive" ones. But as you can tell from my use of quotation marks, I find the notion of intuitive data to be misleading—as the notion of "bias" in data is misleading. Data is never intuitive, just as it's never free of bias. Indeed, one way of describing intuitive data is merely data that perfectly aligns with our most fundamental, unconscious assumptions: our deepest biases.

9. This debate that my students have, in which one group advocates for a common, stable understanding of underlying concepts, and another group proposes that concepts arise only within local contexts, is a familiar one that recurs in various forms. A particularly felicitous set of literature in this regard concerns the negotiation of values in design. For instance, JafariNaimi, Nathan, and Hargraves (2015) contend that values cannot be "identified and applied" as approaches such

as Batya Friedman's Value-Sensitive Design might suggest, but need to be tendered as "hypotheses" through the design process (Friedman, Kahn, and Borning 2006). In the classroom scenario that I describe in this essay, JafariNaimi, Nathan, and Hargraves's work suggests that data designers should not aim to fully understand concepts like race and ethnicity before design begins, but should instead use the design process to explore the intersection between concept and its implementation as data, for a specific situation. Additional work on values in design that adopts a similar orientation includes Binder et al. (2011) on design "things," or conditions for debate, Shilton (2013) on values levers, and LeDantec (2016) on design publics.

10. Kim TallBear (2013) provides an incisive discussion of genetic data and indigeneity.

11. Upon his return home, Gates had not been able to open his front door, and so, with the assistance of his cab driver, he forced it open. Some of Gates's neighbors called the police. Media coverage of Gates's arrest culminated in President Obama's invitation to both Gates and the officer who arrested him to attend a "beer summit" at the White House. See Jen (2009); Cooper (2009); Cooper and Goodnough (2009).

12. In what kind of sociotechnical system might DNA analysis have led to a different outcome in Gates's case? Ironically, perhaps one in which racial identification was controlled by the state under official standards of codification, where one could produce documentation to correct the faulty impressions of law enforcement—once again, we have arrived at a form of data implementation frighteningly similar to that of apartheid in South Africa. Even if our goals for using such data were very different—to rectify inequities rather than enforcing them—this similarity should make us cautious. (Once again, although I am using race and ethnicity as my specific example here, it is only an example. We should be cautious in all our technical decisions regarding data implementation.)

13. Much of the data from Statistics Denmark comes from the Central Person Registry (CPR), which documents all legal residents, each of whom is identified with a unique number (the CPR number). Among other information, the central registry records age, gender, marital status, and current address, along with name and citizenship.

14. As documented in a 2017 European Commission report on the status of equality data collection in EU member states, Denmark's strategy of using country of origin, parents' country of origin, and citizenship as a sort of proxy for more specific race and ethnicity data is common among EU countries (Farkas 2017). At the time of the report, only Finland, Ireland, and the United Kingdom had national requirements to collect data on racial and ethnic minorities, while some member states, such as France, enact barriers to prohibit this type of data collection. Advocacy groups such as the European Network Against Racism (ENAR) generally promote more widespread collection of race and ethnicity data to better understand and address inequities—with the caveat that such data collection should be anonymous and confidential (ENAR 2015).

15. These aids often involve material arrangements and supplements (e.g., instruments, recorded information). In the kitchen, the "spoons" are often clustered together in their own slot in a cutlery organizer, for instance, while the "mixing spoons" might cohabit with the spatula, tongs, and other "utensils" in another location. Hutchins (1995) describes this empirically as distributed cognition, and Barad (2007) provides a philosophical derivation via her theory of agential realism.

16. The World Meteorological Association still maintains a cloud atlas with accompanying observational instructions (https://cloudatlas.wmo.int/en/home.html).

17. A few recent examples of such studies include Edwards et al. (2011); Muller et al. (2019); Pine and Liboiron (2015); Tanweer, Fiore-Gartland, and Aragon (2016); and Vertesi and Dourish (2011). (Additional references on this topic appear in chapter 4.)

18. The particularities of interpretive flexibility in the context of standards development are discussed by Millerand and Bowker (2009).

19. As examples of scholarship that relies on a conception of classification as an argument or perspective rather than a truth, see, for instance, Andersen (2006), Furner (2010), Hansson (2005), and Mai (2011).

20. Historian Ann Blair (2010) traces this tendency to the early modern period, as bibliographies transitioned from being the unique product of a distinctive scholarly mind (a medieval sensibility) to standardized, comprehensive systems (the emergence of modern "information management"). While medieval authors of compiled quotations emphasized their unique ability to select the best bits from their sources (the "flowers" of a work; medieval compilations were accordingly called *florilegia*), early modern compilers deferred this judgment to the reader (or user). The compiler's claim to scholarship shifts from one of particular discernment in the selection, arrangement, and relation of material to one of painstaking, but relatively "mechanical" labor.

21. Park (2009) reviews metadata quality criteria and determines accuracy, comprehensiveness, and consistency to be the most prevalent. Weagley, Gelches, and Park (2010) provide an example of a data quality assessment using these three criteria. The consistency metric, in particular, has a long history within information science; presumably, if data is consistent, then the data creation protocols have been closely followed, and human judgment has been successfully constrained. But tests of interindexer consistency invariably achieve a lower standard than hoped for (see, for instance, Markey 1984; Olson and Wolfram 2008). In contrast, quality criteria that require judgment to operationalize, such as "fitness for purpose," as discussed by Lee, Clarke, and Perti (2015), appear less frequently.

22. Versions of *Protocols of the Elders of Zion*, which was originally written in Russian, appear in English under various similar titles, including *Protocols of the Wise Men of Zion* and *Protocols of the Meetings of the Learned Elders of Zion*. (For brief historical summaries of the *Protocols* and their reception, see Rothstein [2006]; Anti-Defamation League [n.d.].)

23. In library cataloging, data is created for each variation of a resource, although not every copy of each variation. This means that if a textbook appears in two editions, and the most recent edition has an electronic version and a print version, three records for the textbook appear in the catalog: one for the first edition (only in print), one for the electronic version of the second edition, and one for the print version of the second edition. (It doesn't matter whether the library has one copy, three copies, or twenty copies of the first edition; only one record is created.) Because data is separately created for each variation, each of the three records for the textbook may include different subject headings. In other words, the catalog data may assert that the same textbook is *about* different things, from one edition to the next and even from the print version to the electronic version of the same edition (which probably have exactly the same content). Often, different variations of the same resource have similar subject headings—but sometimes they don't.

24. Cataloging data doesn't display creation or modification dates. The cataloging date can't be before the publication date, but an old resource may enter a library collection at any time, and data is sometimes modified after its initial creation. (In other words, if the publication date is 1920, the subject headings could either be from circa 1920 or from some unknown later date.) When there are many versions to compare, one can make informed guesses about which data may have been updated on the basis of trends in cataloging data over time. Via such a comparison, I think it reasonably likely that the data in table 1 was created close to the publication date.

25. Jonathan Furner (2007) makes a similar argument in advocating for critical race theory (CRT) as a conceptual substrate for a just library service. Furner employs CRT to criticize a decision made by editors of the Dewey Decimal Classification (DDC) to excise "race" as a topic, following scientific consensus on the nonexistence of race as a biological distinction. More recently, in alignment with grassroots initiatives such as We Need Diverse Books and Diverse Book Finder, scholars of librarianship have put forth speculative interventions that explore values-oriented assessment in the catalog (Clarke and Schoonmaker 2020) and reader's advisory (Lawrence 2020). Such practice-oriented proposals have arisen in the context of increasing calls to de-neutralize the curriculum of library and information science (e.g., Pawley 2006; Cooke, Sweeney, and Noble 2016; Gibson, Hughes-Hassell, and Threats 2018.) Anthony Dunbar (2006) and Kelvin White (2017) promote similar initiatives for archives and recordkeeping. Where such proposals differ from those like Patrick Wilson's (1983) and Phil Agre (1995) is that, while Wilson and Agre advocate for human judgment in aboutness data (with Agre's proposal as specifically in contrast to automated retrieval techniques), they both imagine a data creator as a neutrally skeptical presence, without a personal stake in the resulting data or the systems that make use of it.

26. In his elegant exegesis of graph-based data structures, Neal Thomas (2018) describes the situation that I have examined empirically, via the specific example of cataloging aboutness, more generally in these terms:

Beyond the important insight that our collective prejudices and blind spots in thinking risk material instantiation into information infrastructures, we need to be thinking about potential blind spots in the semiotic regimes that produce and maintain the initial biunivocalizing conditions for classification; those logics that connect representative to represented, signifier to signified, individualized self to other, and so on. Reflexivity around classificatory practice matters, but so too does reflexivity around the underlying metaphysical premises of those practices. (Thomas 2018, 133)

27. Birger Hjørland (1992) suggested something similar in proposing that aboutness should be framed as the "epistemic potential" of a work, that is, its potential to figure into knowledge in some way. But Hjørland, similarly to the early-twentieth-century library classificationist Henry Bliss, saw knowledge as converging onto a disciplinary consensus, so that although the epistemic potential of a work may be unknown today, its status will clarify as time proceeds. The generation of this data is still conceived, therefore, as a logically derived process of scientific discovery rather than a form of human expression. More recently, Hauser and Tennis (2019) contribute the notion of an *episemantics* of aboutness, in which the meaning of index terms emerges around their use both as index terms and in general discourse.

28. The Cranfield tests of indexing devices compared the retrieval performance of various methods for expressing the subject matter of documents, including the use of terms drawn from painstakingly created indexing languages as well as the use of terms drawn from the documents themselves (e.g., from the document title) (Cleverdon 1967). The Cranfield tests suggested that the vocabulary control of human-created indexing languages mattered little to the retrieval metrics of precision and recall (discussed in chapter 2). That the situation used for the test involved extremely technical articles about aeronautics (with extremely technical and detailed queries on the part of the simulated users) undoubtedly affected the results a great deal. The retrieval of technical documents by researchers was the originating use case upon which today's understanding of information seeking activities still rests. Bernhard Rieder's (2020) *Engines of Order* describes this milieu well.

29. Bill Maron's initial (1961) experiments for probabilistically identifying textual "clues" as aboutness proxies is explicit that the classificatory algorithm generating retrieval results does not reveal anything about a document's actual subject matter, according to any human understanding of describing what documents are about. These automated techniques are not "thinking" like humans. But if our goal is merely to achieve certain retrieval results, this doesn't matter. As Rieder (2020) insightfully explains, the "algorithmic technique" originated by Maron in the context of document retrieval is now, under the banner of "machine learning," used as the basis for quantifying, sorting, ordering, and ranking anything—not just documents. In all these systems, the automatically extracted "features" used to support a certain outcome (e.g., to route certain e-mail to your spam folder or to identify people accused of crimes who should be denied bail) may or may not have any resemblance to the evidence used by humans to make judgments about spamminess

or bail-worthiness, and the process by which such evidence is weighed is likewise incommensurable between machine and person.

30. Safiya Noble's (2018) *Algorithms of Oppression* documents a similar example as she describes the results of a Google search for "black girls" as being riddled with pornography and other demeaning and derogatory results. Tellingly, although Google had adjusted its results for "black girls" as of 2020—one now gets results like sites for Black Girls Code and Black Girls Rock—searching for various other groups (e.g., Ukrainian girls, Thai girls) will bring up results about mail-order brides and the like. I say "tellingly" because the Google search engine has clearly not assimilated that crudely objectifying women as sexual objects is an inappropriate perspective, that is, the search engine's approximation of *aboutness* has not become more sophisticated or humanlike. Instead of assessing its ranking generally in accordance with human values, Google has adjusted a particular result to avoid public censure.

31. Jackson and Barbrow (2015) make a similar proposal in comparing data collection standards to jazz standards: standards as the basis for informed improvisation, which they feel characterizes actual, as opposed to ideal, practice.

32. This sense of genre arises from work in rhetoric and composition, of which early examples include Miller (1984), Yates (1989), and Bazerman (2000). Andersen (2015) argues for a genre-oriented perspective on information organization, contending that when people make decisions about sorting and ordering, this is a form of communicative action. But similar arguments can arise from various standpoints. Frohmann (1990), for instance, uses a perspective derived from Wittgensteinian language games to understand data creation protocols (in the form of "indexing rules") in a similar fashion.

CONCLUSION

1. An earlier take on Grandma's peanut butter sandwiches appears in Feinberg (2018).

2. This project was developed as a pedagogical intervention, but the first run in 2014 made me realize that it was also interesting from a research perspective, and the 2014 students were quite enthusiastic about this prospect. So, for subsequent classes I received approval from the university Institutional Review Board (IRB) to use student data for research purposes. (For students that consent, I can use their anonymized papers as research data.)

3. Sometimes people think that data created by students must be deficient in some way, and that "real" data would be of better quality. But this is precisely wrong. For one, the students had better qualifications than most professional data creators. These were graduate students in an advanced elective; indeed, many of them were already working in professional capacities. Furthermore, these students spent much more time and effort on every aspect of the data creation process—from the

development of implementation guidelines to the data collection itself—than most professional environments would allow. The student data was uniformly excellent.

4. The Trump administration's disregard for facts was well documented. As of May 2021, Politifact (https://www.politifact.com/) had rated only 12 percent of the statements that it had checked for Donald Trump as True (3 percent) or Mostly True (9 percent). In contrast, 53 percent of Trump's statements were rated as either False (36 percent) or Pants on Fire (17 percent). (Pants on Fire statements make "ridiculous claims," in addition to being false.) When I had checked Politifact six months earlier, in December 2020, 16 percent of Trump's statements were rated as Pants on Fire, but otherwise the accumulated ratings were consistent.

REFERENCES

Acker, Amelia, and Joan Donovan. 2019. Data craft: A theory/methods package for critical Internet studies. *Information, Communication & Society* 22(11): 1590–1609.

Adams, Richard. 2020. Nearly 40 percent of A level result predictions to be downgraded in England. *The Guardian*, August 7, 2020. https://www.theguardian.com/education/2020/aug/07/a-level-result-predictions-to-be-downgraded-england.

Adler, Melissa. 2017. *Cruising the Library: Perversities in the Organization of Knowledge.* New York: Fordham University Press.

Agre, Phil. 1995. Institutional circuitry: Thinking about the forms and uses of information. *Information Technology and Libraries* 14(4): 225–230.

Aitchison, Jean, Alan Gilchrist, and David Bawden. 2003. *Thesaurus Construction and Use: A Practical Manual.* 4th ed. London: Routledge.

Amoore, Louise. 2020. *Cloud Ethics.* Durham, NC: Duke University Press.

Ancestry. 2017. Behind the Ancestry commercial: Livie from all nations. Blog post, January 13, 2017. https://blogs.ancestry.com/cm/behind-the-ancestry-commercial-with-livie/. Video, 30 seconds. https://www.ispot.tv/ad/djTa/ancestrydna-testimonial-livie.

Andersen, Jack. 2006. The public sphere and discursive activities: Information literacy as sociopolitical skills. *Journal of Documentation* 62(2): 213–228.

Andersen, Jack. 2015. Re-describing knowledge organization: A genre and activity-based view. In *Genre Theory in Information Studies*, edited by Jack Andersen and Laura Skouvig, 13–42. Studies in Information, vol. 11. Bingley, UK: Emerald.

Andersen, Jack, and Sille Obelitz Søe. 2020. Communicative actions we live by: The problem with fact-checking, tagging or flagging fake news—the case of Facebook. *European Journal of Communication* 35(2): 126–139.

Andersen, Kristina, and Ron Wakkary. 2019. The Magic Machine workshops: Making personal design knowledge. In *Proceedings of the 2019 CHI Conference on Human Factors in Computing Systems*, Paper 112, 1–13. New York: Association for Computing Machinery.

Anderson, Margo. 2015. *The American Census: A Social History.* 2nd ed. New Haven, CT: Yale University Press.

Andres, Tommy. 2014. Why are sticks of butter long and skinny in the East, and short and fat in the West? *Marketplace Weekend*, October 31, 2014. https://www.marketplace.org/2014/10/31/why-are-sticks-butter-long-and-skinny-east-short-and-fat-west/.

Angwin, Julia, Jeff Larson, Surya Mattu, and Lauren Kirchener. 2016. Machine bias. *ProPublica*, May 23, 2016. https://www.propublica.org/article/machine-bias-risk-assessments-in-criminal-sentencing.

Anti-Defamation League. n.d. A hoax of hate: *Protocols of the Learned Elders of Zion*. Accessed May 10, 2021. https://www.adl.org/resources/backgrounders/a-hoax-of-hate-the-protocols-of-the-learned-elders-of-zion.

Anzaldúa, Gloria. 1987. *Borderlands/La Frontera: The New Mestiza*. San Francisco: Aunt Lute Press.

Ashok, Sowmiya. 2016. The rise of American "others." *The Atlantic*, August 27, 2016. https://www.theatlantic.com/politics/archive/2016/08/the-rise-of-the-others/497690/.

Atici, Levent, Sarah Kansa, Justin Lev-Tov, and Eric Kansa. 2012. Other people's data: A demonstration of the imperative of publishing primary data. *Journal of Archaeological Method and Theory* 4(3): 1–19.

Baca, Murtha, Patricia Harpring, Elisa Lansing, Linda McRae, and Anne Whiteside, on behalf of the Visual Resources Association (VRA). 2006. *Cataloging Cultural Objects: a Guide to Describing Cultural Works and Their Images*. Chicago: American Library Association.

Baker, Thomas, Sean Bechhofer, Antoine Isaac, Alistair Miles, Guus Schreiber, and Ed Summers. 2013. Key choices in the design of simple knowledge organization system (SKOS). *Journal of Web Semantics* 20: 35–49.

Balsamo, Luigi. 1990. *Bibliography, History of a Tradition*. Translated by William A. Pettas. Berkeley: Bernard M. Rosenthal.

Barad, Karen. 2007. *Meeting the Universe Halfway: Quantum Physics and the Entanglement of Matter and Meaning*. Durham, NC: Duke University Press.

Bardzell, Jeff. 2011. Interaction criticism: An introduction to the practice. *Interacting with Computers* 23(6): 604–621.

Bardzell, Jeff, and Shaowen Bardzell. 2013. What is "critical" about critical design? In *Proceedings of the SIGCHI Conference on Human Factors in Computing Systems (CHI '13)*, 3297–3306. New York: Association for Computing Machinery.

Bardzell, Jeff, and Shaowen Bardzell. 2015. *Humanistic HCI*. Synthesis Lectures on Human-Centered Informatics. Kentfield, CA: Morgan & Claypool. https://doi.org/10.2200/S00664ED1V01Y201508HCI031.

Bates, Jo, Yu-Wei Lin, and Paula Goodale. 2016. Data journeys: Capturing the socio-material constitution of data objects and flows. *Big Data and Society* 3(2).

Bates, Marcia. 1976. Rigorous systematic bibliography. *RQ* 16: 5–24.

Bates, Marcia. 1986. What is a reference book? *RQ* 26(1): 37–57.

Bates, Marcia. 1989. The design of browsing and berrypicking techniques for the online search interface. *Online Review* 13(5).

Bawden, David, and Lyn Robinson. 2016. Information and the gaining of understanding. *Journal of Information Science* 42(3): 94–299.

Bazerman, Charles. 2000. *Shaping Written Knowledge: the Genre and Activity of the Experimental Article in Science*. Madison: University of Wisconsin Press.

Beghtol, Clare. 1986. Semantic validity: Concepts of warrant in bibliographic classification systems. *Library Resources and Technical Services* 30(2): 109–123.

Beghtol, Clare. 2001. Relationships in classificatory structure and meaning. In *Relationships in the Organization of Knowledge*, edited by Carol Bean and Rebecca Green, 99–113. Dordrecht, Netherlands: Kluwer Academic.

Beghtol, Clare. 2002. A proposed ethical warrant for global knowledge representation and organization systems. *Journal of Documentation* 58(5): 507–532.

Bellacasa, María Puig de la. 2017. *Matters of Care: Speculative Ethics in More than Human Worlds*. Minneapolis: University of Minnesota Press.

Bender, Emily, and Alexander Koller. 2020. Climbing towards NLU: On meaning, form, and understanding in the age of data. In *Proceedings of the 58th Annual Meeting for Computational Linguistics:* 5185–5198. n.p.: Association for Computational Linguistics.

Benjamin, Ruha. 2019. *Race After Technology: Abolitionist Tools for the New Jim Code*. Cambridge, UK: Polity.

Benjamin, Walter. 1999. *The Arcades Project*. Translated by Howard Eiland and Kevin McLaughlin. Cambridge, MA: Belknap Press.

Berman, Sanford. 1971. *Prejudices and Antipathies: A Tract on the LC Heads Concerning People*. Metuchen, NJ: Scarecrow Press.

Besterman, Theodore. 1936. *The Beginnings of Systematic Bibliography*. Oxford, UK: Oxford University Press.

Bildt, Carl. 2018. The world needs to pay attention to the crisis over Macedonia's name. *Washington Post*, February 5, 2018. https://www.washingtonpost.com/news/global-opinions/wp/2018/02/05/the-world-needs-to-pay-attention-to-the-crisis-over-macedonias-name/.

Binder, Thomas, Giorgio De Michelis, Pelle Ehn, Giulio Jacucci, Per Linde and Ina Wagner. 2011. *Design Things*. Cambridge, MA: MIT Press.

Blair, Ann. 2010. *Too Much to Know: Managing Scholarly Information Before the Modern Age*. New Haven, CT: Yale University Press.

Bliss, Henry Evelyn. 1929. *The Organization of Knowledge and the System of the Sciences*. New York: Henry Holt.

Bogost, Ian. 2012. *Alien Phenomenology, or, What It's Like to Be a Thing*. Minneapolis: University of Minnesota Press.

Bowker, Geoffrey, and Susan Leigh Star. 1999. *Sorting Things Out: Classification and Its Consequences*. Cambridge, MA: MIT Press.

Bowker, Geoffrey. 2000. Biodiversity datadiversity. *Social Studies of Science* 30(5): 643–683.

Branswell, Helen. 2020. It's been sequenced. It's spread across borders. Now the new pneumonia-causing virus needs a name. *STAT*. January 23, 2020. https://www.statnews.com/2020/01/23/its-been-sequenced-its-spread-across-borders-now-the-new-pneumonia-causing-virus-needs-a-name/.

Brantley, Ben. 2017. A Hamlet poised between cultures and languages. *New York Times*, May 22, 2017. https://www.nytimes.com/2017/05/22/theater/hamlet-review-sheen-center.html.

Broughton, Vanda. 2006. *Essential Thesaurus Construction*. London: Facet.

Broussard, Meredith. 2020. When algorithms give real students imaginary grades. *New York Times*, September 8, 2020. https://www.nytimes.com/2020/09/08/opinion/international-baccalaureate-algorithm-grades.html.

Bryant, Rebecca. 2000. *Discovery and Decision: Exploring the Metaphysics and Epistemology of Scientific Classification*. Madison, NJ: Farleigh Dickinson University Press.

Buckland, Michael. 1991. Information as thing. *Journal for the American Society of Information Science* 42(5): 351–360.

Buckland, Michael. 1997. What is a "document"? *Journal for the American Society of Information Science* 48(9): 804–809.

Buckland, Michael. 2012. Obsolescence in subject description. *Journal of Documentation* 68(2): 154–161.

Bullard, Julia. 2017. Warrant as a means to study classification system design. *Journal of Documentation* 73(1): 75–90.

Bullard, Julia. 2018. Curated folksonomies: three implementations of structure through human judgment. *Knowledge Organization* 45(8): 643–652.

Capra, Rob, Jaime Arguello, Anita Crescenzi, and Emily Vardell. 2015. Differences in the use of search assistance for tasks of varying complexity. In *Proceedings of the 38th International ACM SIGIR Conference on Research and Development in Information Retrieval*, 23–32. New York: Association for Computing Machinery.

Carlyle, Allyson. 2015. The policeman's beard was what? Representation and reality in knowledge organization and description. In *Proceedings of the iConference 2015*. Urbana: University of Illinois. http://hdl.handle.net/2142/73642.

Carter, Daniel, Amelia Acker, and Daniel Sholler. 2021. Investigative approaches to researching information technology companies. *Journal of the Association for Information Science and Technology* 72(6): 655–666.

Cartwright, Nancy. 1991. Replicability, reproducibility, and robustness: Comments on Harry Collins. *History of Political Economy* 23(1): 143–155.

Caswell, Michelle. 2012. Using classification to convict the Khmer Rouge. *Journal of Documentation* 68 (2): 162–184.

Caswell, Michelle. 2014. *Archiving the Unspeakable: Silence, Memory, and the Photographic Record in Cambodia.* Madison: University of Wisconsin Press.

Caswell, Michelle, Ricardo Punzalan, and T-Kay Sangwand. 2017. Critical archival studies: An introduction. *Journal of Critical Library and Information Studies* 1(2). https://doi.org/10.24242/jclis.v1i2.50.

Chan, Lois Mai and Athena Salaba. 2016. *Cataloging and Classification: An Introduction.* 4th ed. Lantham, MD: Rowman and Littlefield.

Chang, Wen-Wei, Elisa Giaccardi, Lin-Lin Chen, and Rung-Huei Liang. 2017. "Interview with things": A first-thing perspective to understand the scooter's everyday socio-material network in Taiwan. In *Proceedings of the 2017 Conference on Designing Interactive Systems (DIS '17)*: 1001–1012. New York: Association for Computing Machinery.

Chatman, Elfreda. 1991. Life in a small world: Application of gratification theory to information-seeking behavior. *Journal of the American Society for Information Science* 42(6): 438–449.

Chatman, Elfreda. 1996. The impoverished life-world of outsiders. *Journal of the American Society for Information Science* 47(3): 193–206.

Chatman, Elfreda. 1999. A theory of life in the round. *Journal of the American Society for Information Science* 50(3): 207–217.

Chin, Elizabeth. 2016. *My Life with Things: The Consumer Diaries.* Durham, NC: Duke University Press.

Cifor, Marika. 2016. Aligning bodies: Collecting, arranging, and describing hatred for a critical queer archives. *Library Trends* 64(4): 756–775.

Cifor, Marika, Patricia Garcia, T. L. Cowan, Jasmine Rault, Tonia Sutherland, Anita Say Chan, Jennifer Rode, Anna Lauren Hoffman, Niloufar Salehi, and Lisa Nakamura. 2019. Feminist Data Manifest-No. https://www.manifestno.com/.

Clarke, Rachel Ivy, and Sayward Schoonmaker. 2020. The critical catalog: Library information systems, tricksterism, and social justice. In *Proceedings of the 2020 ACM Conference on Human Factors in Computing Systems (CHI '20)*, 1–13. New York: Association for Computing Machinery.

Classification Research Group. 1955. The need for a faceted classification as the basis for all methods of information retrieval. *Library Association Record* 57(7): 262–268.

Clement, Tanya. 2016. Towards a rationale of audio-text. *Digital Humanities Quarterly* 10(2). http://www.digitalhumanities.org/dhq/vol/10/3/000254/000254.html.

Cleverdon, Cyril. 1967. The Cranfield tests on index language devices. Reprinted in *Readings in Information Retrieval*, 1997, edited by Karen Spärck Jones and Peter Willet, 47–59. New York: Morgan Kaufman.

Collins, Harry M. 1985. *Changing Order: Replication and Induction in Scientific Practice*. London: Sage.

Cook, Terry. 1997. What is past is prologue: A history of archival ideas since 1898, and the future paradigm shift. *Archivaria* 43: 17–63.

Cooke, Nicole A., Miriam E. Sweeney, and Safiya Umoja Noble. 2016. Social justice as topic and tool: An attempt to transform an LIS curriculum and culture. *Library Quarterly* 86(1): 107–24.

Cooper, Helene. 2009. Obama criticizes arrest of Harvard professor. *New York Times* July 22, 2009. https://www.nytimes.com/2009/07/23/us/politics/23gates.html.

Cooper, Helene, and Abby Goodnough. 2009. Over beers, no apologies, but plans to have lunch. *New York Times* July 30, 2009. https://www.nytimes.com/2009/07/31/us/politics/31obama.html.

Costanza-Chock, Sasha. 2020. *Design Justice: Community-Led Practices to Build the Worlds We Need*. Cambridge, MA: MIT Press.

Couldry, Nick, and Ulises Mejias. 2019. *The Costs of Connection: How Data Is Colonizing Human Life and Appropriating It for Capitalism*. Stanford, CA: Stanford University Press.

Crescenzi, Anita, Austin Ward, Yuan Li, and Rob Capra. 2021. Supporting metacognition in exploratory search with the OrgBox. In *Proceedings of the 44th International ACM SIGIR Conference on Research and Development in Information Retrieval (SIGIR '21)*, 1197–1207. New York: Association for Computing Machinery.

Currie, Morgan, Britt Paris, Irene Pasquetto, and Jennifer Pierre. 2016. The conundrum of police-officer-involved homicides: Counter-data in Los Angeles county. *Big Data and Society* 3(2).

Cutter, Charles. 1904. *Rules for a Dictionary Catalog*. 4th ed., rewritten. Washington, DC: Government Printing Office.

Daston, Lorraine. 2015. Cloud physiognomy: Describing the indescribable. *Representations* 135 (Summer 2015): 45–71.

Daston, Lorraine, and Peter Galison. 2007. *Objectivity*. New York: Zone Books.

Day, Ron. 2015. *Indexing It All: The Subject in the Age of Documentation, Information, and Data*. Cambridge, MA: MIT Press.

Day, Ron. 2019. *Documentarity: Evidence, Ontology, and Inscription*. Cambridge, MA: MIT Press.

Desmond-Harris, Jenée. 2021. My cousin turned my family against me over a baby name. *Slate*, July 1, 2021. https://slate.com/human-interest/2021/07/dear-prudence-cousin-stillborn-baby-name-stolen.html.

D'Ignazio, Catherine, and Lauren Klein. 2020. *Data Feminism*. Cambridge, MA: MIT Press.

Dominus, Susan. 2017. When the revolution came for Amy Cuddy. *The New York Times Magazine*, October 18, 2017. https://www.nytimes.com/2017/10/18/magazine/when-the-revolution-came-for-amy-cuddy.html.

Domonoske, Camilla. 2017a. Sheila Michaels, who helped bring honorific "Ms" to the masses, dies at 78. *NPR*, July 7, 2017. https://www.npr.org/sections/thetwo-way/2017/07/07/535978012/sheila-michaels-who-helped-bring-honorific-ms-to-the-masses-dies-at-78.

Domonoske, Camilla. 2017b. When "Miss" meant so much more: How one woman fought Alabama—and won. *NPR*. Code Switch podcast, November 30, 2017. https://www.npr.org/sections/codeswitch/2017/11/30/567177501/when-miss-meant-so-much-more-how-one-woman-fought-alabama-and-won.

Doty, Mark. 2002. *Still Life with Oysters and Lemon*. Boston: Beacon Press.

Dourish, Paul. 2014. NoSQL: The shifting materialities of database technology. *Computational Culture* 4.

Dourish, Paul, and Edgar Gomez Cruz. 2018. Datafication and data fiction: Narrating with big data. *Big Data and Society* 5(2).

Dourish, Paul, and Melissa Mazmanian. 2013. Media as material: Information representations as material foundations for organizational practice. In *How Matter Matters: Objects, Artifacts, and Materiality in Organization Studies*, edited by Paul R. Carlile, Davide Nicolini, Anne Langley, and Haridmos Tsoukas. Oxford, UK: Oxford University Press.

DR (Danish Radio). 2019. Prisvinder kritiserer statsminister in takketale. (Prize winner criticizes prime minister in thank-you speech.) October 29, 2019. https://www.dr.dk/nyheder/kultur/boeger/prisvinder-kritiserer-statsminister-i-takketale.

Drabinski, Emily. 2013. Queering the catalog: Queer theory and the politics of correction. *Library Quarterly* 83(2): 94–111.

Drucker, Johanna. 2006. Graphical readings and the visual aesthetics of textuality. *Text* 16: 267–276.

Drucker, Johanna. 2013. Performative materiality and theoretical approaches to interface. Digital *Humanities Quarterly* 7(1). http://www.digitalhumanities.org/dhq/vol/7/1/000143/000143.html.

Duff, Wendy, and Verne Harris. 2002. Stories and names: Archival description as narrating records and constructing meanings. *Archival Science* 2: 263–285.

Duhigg, Charles. 2017. Outcry over EpiPen prices hasn't made them lower. *New York Times*, June 4, 2017. https://www.nytimes.com/2017/06/04/business/angry-about-epipen-prices-executive-dont-care-much.html.

Dunbar, Anthony W. 2006. Introducing critical race theory to archival discourse: getting the conversation started *Archival Science* 6: 109–129.

Dupré, John. 1993. *The Disorder of Things: Metaphysical Foundations of the Disunity of Science*. Cambridge, MA: Harvard University Press.

Dupré, John. 2006. Scientific classification. *Theory, Culture, and Society* 23(2–3): 30–33.

Ebeling, Mary F. E. 2016. *Healthcare and Big Data: Digital Specters and Phantom Objects*. New York: Palgrave Macmillan.

Eco, Umberto. 2009. *The Infinity of Lists*. New York: Rizzoli.

Edwards, Paul, Geoffrey Bowker, Steven Jackson, and Robin Williams. 2009. An agenda for infrastructure studies. *Journal of the Association for Information Systems* 10(5): Article 6.

Edwards, Paul, Matthew Mayernik, Archer Batcheller, Geoffrey Bowker, and Christine Borgman. 2011. Science friction: Data, metadata, and collaboration. *Social Studies of Science* 41(5): 667–690.

Eggert, Paul. 1999. Where are we now with authorship and the work? *Yearbook of English Studies* 88–96.

Eggert, Paul. 2019. *The Work and the Reader in Literary Studies: Scholarly Editing and Book History*. Cambridge, UK: Cambridge University Press.

Eika, Jonas. 2019. Speech on receiving the Nordic Council Literature Prize 2019. Text published in *POV International*. https://pov.international/jonas-eika-tale-ved-modtagelsen-af-nordisk-rads-litteraturpris-2019/.

Elings, Mary, and Gunter Weibel. 2007. Metadata for all: Descriptive standards and metadata sharing across libraries, archives, and museums. *First Monday* 12(3). http://firstmonday.org/article/view/1628/1543.

Ensmenger, Nathan. 2010. *The Computer Boys Take Over: Computers, Programmers, and the Politics of Technical Expertise*. Cambridge, MA: MIT Press.

Erdelez, Sanda. 1997. Information encountering: a conceptual framework for accidental information discovery. In *ISIC '96: Proceedings of an International Conference on Information Seeking in Context*, 412–421. London: Taylor Graham.

Erdelez, Sanda. 2005. Information encountering. In *Theories of Information Behavior*, edited by Karen Fisher, Sanda Erdelez, and Lynn McKechnie, 179–184. Medford, NJ: Information Today.

Ereshefsky, Marc. 2007. *The Poverty of the Linnaean Hierarchy: A Philosophical Study of Biological Taxonomy*. Cambridge, UK: Cambridge University Press.

European Network Against Racism (ENAR). 2015. *Equality Data Collection: Facts and Principles*. https://www.enar-eu.org/IMG/pdf/edc-general_factsheet_final.pdf.

Faniel, Ixchel M., and Ann Zimmerman. 2011. Beyond the data deluge: A research agenda for large-scale data sharing and reuse. *International Journal of Digital Curation* 6(1): 58–69.

Farkas, Lilla. 2017. *Data Collection in the Field of Ethnicity: Analysis and Comparative Review of Equality Data Collection Practices in the European Union*. European Commission, Directorate-General for Justice and Consumers, Directorate D-Equality. https://doi.org/10.2838/447194.

Feinberg, Melanie. 2013. Beyond digital and physical objects: The intellectual work as a concept of interest for HCI. In *Proceedings of the ACM SIGCHI Conference on Human Factors in Computing Systems—CHI 2013*, 3317–3326. New York: Association for Computing Machinery.

Feinberg, Melanie. 2016. The value of discernment: Making use of interpretive flexibility in metadata generation and aggregation. *Information Research* 22(1). http://www.informationr.net/ir/22-1/colis/colis1649.html.

Feinberg, Melanie. 2017a. A design perspective on data. In *Proceedings of the ACM Conference on Human Factors in Computing Systems—CHI 2017*, 2952–2963. New York: Association for Computing Machinery.

Feinberg, Melanie. 2017b. Material vision. In *Proceedings of the ACM Conference on Computer Supported Cooperative Work and Social Computing 2017 (CSCW 2017)*, 604–617. New York: Association for Computing Machinery.

Feinberg, Melanie. 2018. Factotem: What is information access for? *Cataloging and Classification Quarterly* 56(8): 665–682.

Feinberg, Melanie, Ramona Broussard, and Eryn Whitworth. 2016. Framing a set: Understanding the curatorial character of personal digital bibliographies. *Interacting With Computers* 28(1): 102–124.

Feinberg, Melanie, Daniel Carter, Julia Bullard, and Ayse Gursoy. 2017. Translating texture: Design as integration. In *Proceedings of the ACM Conference on Designing Interactive Systems—DIS 2017*, 297–307. New York: Association for Computing Machinery.

Feinberg, Melanie, Will Sutherland, Mohammad Hossein Jarrahi, Sarah Beth Nelson, and Arcot Rajasekar. 2020. The new reality of replicability: The role of data work in scientific research. In *Proceedings of the Association for Computing Machinery (ACM) on Human-Computer Interaction* 4(CSCW1): Article 35.

Fennessy, Julian, et al. 2016. Multi-locus analyses reveal four giraffe species instead of one. *Current Biology* 26(18): 2543–2549.

Floridi, Luciano. 2010. *Information: A Very Short Introduction*. Oxford, UK: Oxford University Press.

Foskett, D. J. 1974. *Classification and Indexing in the Social Sciences.* 2nd ed. London: Butterworths.

Foster, Edward, and David Ellis. 2014. Serendipity and its study. *Journal of Documentation* 70(6): 1015–1038.

Fox, Melodie. 2016. Legal discourse's epistemic interplay with sex and gender classification in the Dewey Decimal classification system. *Library Trends* 64(4): 687–713.

Freeman, Jo. 1975/2013. The tyranny of structurelessness. *WSQ: Women's Studies Quarterly* 41(3): 231–246.

Friedman, Batya, Peter Kahn, and Alan Borning. 2006. Value sensitive design and information systems. In *Human–Computer Interaction and Management Information Systems: Foundations,* edited by Ping Zhang and Dennis Galletta, 348–72. Armonk, NY: M.E. Sharpe.

Frohmann, Bernd. 1990. Rules of indexing: a critique of mentalism in information retrieval theory. *Journal of Documentation* 46(2): 81–101.

Fuller, Thomas. 2014. You call this Thai food? The robotic taster will be the judge. *New York Times* September 28, 2014. https://www.nytimes.com/2014/09/29/world/asia/bad-thai-food-enter-a-robot-taster.html.

Fuller, Thomas. 2021. Daylight attack on two Asian women in San Francisco increases fears. *New York Times,* May 5, 2021. https://www.nytimes.com/2021/05/05/us/asian-attack-san-francisco.html.

Furnas, George, Thomas Landauer, Louis Gomez, and Susan Dumais. 1987. The vocabulary problem in human-system communication. *Communications of the ACM* 30(11): 964–971.

Furner, Jonathan. 2007. Dewey deracialized: A critical race-theoretic perspective. *Knowledge Organization* 24(3): 144–168.

Furner, Jonathan. 2010. Philosophy and information studies. *Annual Review of Information Science and Technology* 44: 161–200.

Furner, Jonathan. 2012. FRSAD and the ontology of subjects of works. *Cataloging and Classification Quarterly* 50: 494–516.

Georgievski, Boris. 2020. Bulgaria asks EU to stop "fake" Macedonian identity. *DW.* September 23, 2020. https://www.dw.com/en/bulgaria-asks-eu-to-stop-fake-macedonian-identity/a-55020781.

Gibson, Amelia, Sandra Hughes-Hassell, and Megan Threats. 2018. Critical race theory in the LIS curriculum. In *Re-envisioning the MLS: Perspectives on the Future of Library and Information Science Education,* 49–70. Advances in Librarianship, vol. 44B. Bingley, UK: Emerald.

Gibson, Amelia, and John Martin. 2019. Re-situating information poverty: information marginalization and parents of individuals with disabilities. *Journal of the Association for Information Science and Technology* 70(5): 476–487.

Gilbert, Jeremy. 2014. *Common Ground: Democracy and Collectivity in an Age of Individualism*. London: Pluto Press.

Gilliland, Anne. 2000. *Enduring Paradigm, New Opportunities: The Value of the Archival Perspective in the Digital Environment*. Council of Library and Information Resources (CLIR). http://www.clir.org/pubs/reports/pub89/contents.html.

Gitelman, Lisa, ed. 2013. *Raw Data Is an Oxymoron*. Cambridge, MA: MIT Press.

Given, Lisa M., and Lianne McTavish. 2010. What's old is new again: The reconvergence of libraries, museums and archives in the digital age. *Library Quarterly* 80(1): 7–32.

Gleick, James. 2011. *The Information: A History, a Theory, a Flood*. New York: Pantheon.

Glushko, Robert J., ed. 2013. *The Discipline of Organizing*. Cambridge, MA: MIT Press.

Goodman, Steven N., Daniele Fanelli, and John P. A. Ioannidis. 2016. What does research reproducibility mean? *Science Translational Medicine* 8(341): 341ps12.

Goodwin, Charles. 1994. Professional vision. *American Anthropologist* 96(3): 606–633.

Gorichanaz, Tim, and Kiersten Latham. 2016. Document phenomenology: A framework for holistic analysis. *Journal of Documentation* 72(6): 1114–1133.

Gulick, Sidney. 1914. *The American Japanese Problem: A Study on the Racial Relations of the East and West*. New York: Scribner.

Haider, Jutta, and Olof Sundin. 2019. *Invisible Search and Online Search Engines*. New York: Routledge.

Haimson, Oliver, and Anna Lauren Hoffman. 2016. Constructing and enforcing "authentic identity" online: Facebook, real names, and non-normative identities. *First Monday*. https://firstmonday.org/ojs/index.php/fm/article/view/6791/5521.

Hansson, Joacim. 2005. Hermeneutics as a bridge between the modern and the postmodern in library and information science. *Journal of Documentation* 61(1): 102–113.

Haraway, Donna. 1988. Situated knowledges: The science question in feminism and the privilege of partial perspective. *Feminist Studies* 14(3): 575–599.

Harding, Sandra, ed. 2004. *The Feminist Standpoint Theory Reader: Intellectual and Political Controversies*. New York: Routledge.

Hartman, Saidiya. 2007. *Lose Your Mother: A Journey Along the Atlantic Slave Route*. New York: Farrar, Straus, and Giroux.

Hauser, Elliott, and Joseph Tennis. 2019. Episemantics: Aboutness as aroundness. *Proceedings of the North American Society for Knowledge Organization 2019*, 27–34.

Hayles, N. Katherine. 2014. Speculative aesthetics and object-oriented inquiry. *Speculations: A Journal of Speculative Realism*: 158–179.

Hjørland, Birger. 1992. The concept of subject in information science. *Journal of Documentation* 48(2): 172–200.

Hjørland, Birger. 2002. Domain analysis in information science: Eleven approaches–traditional as well as innovative. *Journal of Documentation* 58(4): 422–462.

Hjørland, Birger. 2004. Arguments for philosophical realism in library and information science. *Library Trends* 52(3): 488–506.

Hjørland, Birger, and Hanne Albrechtsen. 1995. Toward a new horizon in information science: Domain-analysis. *Journal for the American Society of Information Science* 46(6): 400–425.

Hochschild, Jennifer, Vesla Weaver, and Traci Burch. 2012. *Creating a New Racial Order: How Immigration, Multiracialism, Genomics, and the Young Can Remake Race in America*. Princeton, NJ: Princeton University Press.

Hoffmann, Anna Lauren. 2017. Data, technology, and gender: Thinking about (and from) trans lives. In *Spaces for the Future: A Companion to Philosophy of Technology*, edited by Joseph Pitt and Ashley Shew, 3–13. London: Routledge.

Hoffman, Anna Lauren. 2020. Terms of inclusion: Data, discourse, violence. *New Media and Society* 20(3): 1112–1130.

Hogan, Kristen. 2010. "Breaking secrets" in the catalog: proposing the Black Queer Studies Collection at the University of Texas at Austin. *Progressive Librarian* 34/35: 50–57.

Hong, Nicole, Juliana Kim, Ali Watkins, and Ashley Southall. 2021. Brutal attack on Filipino woman sparks outrage: "Everybody is on edge." *New York Times*, March 30, 2021. https://www.nytimes.com/2021/03/30/nyregion/asian-attack-nyc.html.

Houser, Heather. 2020. *Infowhelm: Environmental Art and Literature in an Age of Data*. New York: Columbia University Press.

Hulme, E. W. 1911. Principles of book classification. *Library Association Record* 13: 354–358, 389–394, and 444–449.

Huvila, Isto. 2016. Awkwardness of becoming a boundary object: Mangle and materialities of reports, documentation data and the archaeological work. *The Information Society* 32(4): 280–297.

Huvila, Isto. 2019. Authoring social reality with documents: from authorship of documents and documentary boundary objects to practical authorship. *Journal of Documentation* 75(1): 44–61.

Hutchins, Ed. 1995. *Cognition in the Wild*. Cambridge, MA: MIT Press.

Ingold, Tim. 2007. *Lines: A Brief History*. London: Routledge.

International Federation of Library Associations (IFLA). 1996. Functional Requirements for Bibliographic Records final report. http://www.ifla.org/VII/s13/frbr/frbr.pdf.

International Standards Organization (ISO). 2011. ISO 25964–1:2011. Thesauri and interoperability with other vocabularies. Part 1: Thesauri and information retrieval.

Irani, Lilly. 2019. "Design thinking": Defending Silicon Valley at the apex of global labor hierarchies. *Catalyst: Feminism, Theory, Technoscience* 4(1): 1–19.

Jackson, Steven J., and Sarah Barbrow. 2015. Standards and/as innovation: protocols, creativity, and interactive systems development in ecology. In *Proceedings of the 33rd Annual ACM Conference on Human Factors in Computing Systems (CHI '15)*, 1769–1778. New York: Association for Computing Machinery.

Jacob, Elin. 1991. Classification and categorization: Drawing the line. In *Proceedings of the 2nd Annual ASIS SIG/CR Classification Workshop*, 63–80. New York: Association for Computing Machinery. http://dx.doi.org/10.7152/acro.v2i1.12548.

JafariNaimi, Nassim, Lisa Nathan, and Ian Hargraves. 2015. Values as hypotheses: Design, inquiry, and the service of values. *Design Issues* 31(4): 91–104.

Jen, Tracy. 2009. Racial talk swirls with Gates arrest. July 21, 2009. *Boston Globe*. http://archive.boston.com/news/education/higher/articles/2009/07/21/racial_talk_swirls_with_gates_arrest/

Joint Steering Committee (JSC) for the Development of Resource Description and Access (RDA). 2009. Statement of objectives and principles for RDA. http://www.rda-jsc.org/archivedsite/docs/5rda-objectivesrev3.pdf.

Kansa, Eric, Sarah Kansa, and Benjamin Arbuckle. 2014. Publishing and pushing: Mixing models for communicating research data in archeology. *International Journal of Digital Curation* 9(1): 57–70.

Kaplan, Sheila. 2018. F.D.A. approves generic EpiPen that may be cheaper. *New York Times*, August 16, 2018. https://www.nytimes.com/2018/08/16/health/epipen-generic-drug-prices.html.

Kay, Eve. 2007. Call me Ms. *The Guardian*. June 29, 2007. https://www.theguardian.com/world/2007/jun/29/gender.uk.

Kelly, Diane. 2009. Methods for evaluating interactive information retrieval systems with users. *Foundations and Trends in Information Retrieval* 3(1–2).

Keyes, Os. 2020. (Mis)gendering. In *Uncertain Archives: Critical Keywords for the Age of Big Data*, edited by Nanna Bonde Thylstrup, Daniela Agostinho, Catherine D'Ignazio, Annie Ring, and Kristin Veel, 339–346. Cambridge, MA: MIT Press.

King, William Davies. 2008. *Collections of Nothing*. Chicago: University of Chicago Press.

Kirk, David, and Abigail Sellen. 2010. On human remains: Values and practice in the home archiving of cherished objects. *ACM Transactions on Computer-Human Interaction* 17(3): 1–43.

Kirschenbaum, Matthew. 2008. *Mechanisms: New Media and the Forensic Imagination*. Cambridge, MA: MIT Press.

Kitsantonis, Niki. 2018. Greeks protest over neighbor's use of the name Macedonia. *New York Times*, January 21, 2018. https://www.nytimes.com/2018/01/21/world/europe/greece-macedonia.html.

Kitzes, Justin, Daniel Turek, and Fatma Deniz. 2017. *The Practice of Reproducible Research: Case Studies and Lessons from the Data-Intensive Sciences*. Berkeley: University of California Press.

Kjellman, Ulrika. 2017. Images as scientific documents in Swedish "race biology": Two practices. *Information Research* 22(1). http://InformationR.net/ir/22-1/colis/colis1655.html.

Kohn, Eduardo. 2013. *How Forests Think: Toward an Anthropology Beyond the Human*. Berkeley: University of California Press.

Koopman, Colin. 2019. *How We Became Our Data: A Genealogy of the Informational Person*. Chicago: University of Chicago Press.

Kwasnik, Barbara. 1999. The role of classification in knowledge representation and discovery. *Library Trends* 48(1): 22–47.

Kyle, Barbara. 1958. Towards a classification for social science literature. *American Documentation* 9(1–4): 168–183.

Lakoff, George. 1987. *Women, Fire, and Dangerous Things*. Chicago: University of Chicago Press.

Lakoff, George, and Mark Johnson. 1980. *Metaphors We Live By*. Chicago: University of Chicago Press.

Langridge, D. W. 1976. *Classification and Indexing in the Humanities*. London: Butterworths.

Latour, Bruno. 1988. Mixing humans with non-humans: The sociology of a door closer. *Social Problems* 35: 298–310.

Latour, Bruno. 1999. *Pandora's Hope: Essays on the Reality of Science Studies*. Cambridge, MA: Harvard University Press.

Law, John. 1984. On the methods of long-distance control: Vessels, navigation and the Portuguese route to India. *The Sociological Review* 32: 234–263.

Law, John. 1992. Notes on the theory of the actor-network: Ordering, strategy, and heterogeneity. *Systems Practice* 5(4): 379–393.

Law, John, and Marianne Elisabeth Lien. 2012. Slippery: Field notes in empirical ontology. *Social Studies of Science* 43(3): 363–378.

Lawrence, E. E. 2020. On the problem of oppressive tastes in the library. *Journal of Documentation* 76(5): 1091–1107.

LeDantec, Christopher. 2016. *Designing Publics*. Cambridge, MA: MIT Press.

Lee, Erika. 2015. *The Making of Asian America: A History*. New York: Simon and Schuster.

Lee, Hur-Li. 2012. Epistemic foundation of bibliographic classification in early China: A Ru classicist perspective. *Journal of Documentation* 68(3): 378–401.

Lee, Jin Ha, Hyerim Cho, Violet Fox, and Andrew Perti. 2013. User-centered approach in creating a metadata schema for video games and interactive media. In *Proceedings of the 13th ACM/IEEE-CS Joint Conference on Digital Libraries (JCDL)*, 229–238. New York: IEEE Press.

Lee, Jin Ha, Rachel Clarke, and Andrew Perti. 2015. Empirical evaluation of metadata for video games and interactive media. *Journal for the Association of Information Science and Technology (JASIST)* 63(12): 2609–2625.

Leslie, Sarah-Jane. 2013. Essence and natural kinds: When science meets preschooler intuition. *Oxford Studies in Epistemology* 4: 108–166.

Lewis, Paul. 1993. UN compromise lets Macedonia be a member. *New York Times*, April 8, 1993. https://www.nytimes.com/1993/04/08/world/un-compromise-lets-macedonia-be-a-member.html.

Levine, Caroline. 2015. *Forms: Whole, Rhythm, Hierarchy, Network*. Princeton, NJ: Princeton University Press.

Levy, David. 2016. *Scrolling Forward: Making Sense of Documents in the Digital Age*. 2nd ed. New York: Arcade.

Light, Ann, Irina Shklovski, and Alison Powell. 2017. Design for existential crisis. In *Proceedings of the 2017 CHI Conference Extended Abstracts on Human Factors in Computing Systems (CHI EA '17)*: 722–734. New York: Association for Computing Machinery.

Liptak, Adam. 2017. Justices reject two gerrymandered North Carolina districts, citing racial bias. *New York Times*, May 22, 2017. https://www.nytimes.com/2017/05/22/us/politics/supreme-court-north-carolina-congressional-districts.html.

Littletree, Sandra, Miranda Belarde-Lewis, and Marisa Duarte. 2020. Centering relationality: a conceptual model to advance indigenous knowledge organization practices. *Knowledge Organization* 47(5): 410–426.

Liu, Alan. 2018. *Friending the Past: The Sense of History in the Digital Age*. Chicago: University of Chicago Press.

López-Alt, J. Kenji. 2011. The Food Lab: Drinks edition. Is Mexican Coke better? *Serious Eats*, September 2, 2011. https://drinks.seriouseats.com/2011/09/the-food-lab-drinks-edition-is-mexican-coke-better-than-regular-coke-coke-taste-test-coke-vs-mexican-coke.html.

Los Angeles Times. 2015. Girl Scout cookies interactive graphic. https://graphics.latimes.com/girl-scout-cookies/.

Loukissas, Yanni. 2019. *All Data Are Local: Thinking Critically in a Data-Driven Society*. Cambridge, MA: MIT Press.

Lubetzky, Seymour. 1969. *Principles of cataloging: Final report. Part 1: Descriptive cataloging*. Institute of Library Research, University of California, Los Angeles. https://eric.ed.gov/?id=ED031273.

Lupton, Deborah. 2016. *The Quantified Self*. Malden, MA: Polity.

Macaluso, Benoit, Vincent Lariviere, Thomas Sugimoto, and Cassidy Sugimoto. 2016. Is science built on the shoulders of women? A study of gender differences in contributorship. *Academic Medicine* 91(8): 1136–1142.

Machlup, Fritz, and Una Mansfield. 1983. Cultural diversity in studies of information. In *The Study of Information: Interdisciplinary Messages*, edited by Fritz Machlup and Una Mansfield. New York: Wiley.

Mai, Jens-Erik. 2011. The modernity of classification. *Journal of Documentation* 67(4): 710–730.

Mak, Bonnie. 2011. *How the Page Matters*. Toronto: University of Toronto Press.

March, Laura, and Sayamindu Dasgupta. 2020. Wikipedia edit-a-thons as sites of public pedagogy. *Proceedings of the Association for Computing Machinery (ACM) on Human-Computer Interaction* 4(CSCW2): Article 100.

Marchionini, Gary. 1995. *Information Seeking in Electronic Environments*. Cambridge, UK: Cambridge University Press.

Markey, Karen. 1984. Interindexer consistency tests: a literature review and report of a test of consistency in indexing visual materials. *Library and Information Science Research* 6: 155–77.

Maron, M. E. 1961. Automatic indexing: an experimental inquiry. *Journal of the ACM* 8(3): 404–417.

Marshall, Brianna, and Marijel (Maggie) Melo. 2020. From needs analysis to power analysis: A framework to examine and broker power in makerspaces. In *Re-Making the Library Makerspace: Critical Theories, Reflections, and Practices*, edited by Marijel (Maggie) Melo and Jennifer Nichols, 83–97. Sacramento, CA: Library Juice Press.

Marshall, Cathy. 2018. Biography, ephemera, and the future of social media archiving. In *Proceedings of the 18th ACM/IEEE Joint Conference on Digital Libraries (JCDL '18)*, 253–262. New York: IEEE Press.

Marshall, Cathy, and Siân Lindley. 2014. Searching for myself: motivations and strategies for self-search. In *Proceedings of the SIGCHI Conference on Human Factors in Computing Systems (CHI '14)*, 3675–3684. New York: Association for Computing Machinery.

Mayernik, Matthew. 2019. Metadata accounts: Achieving data and evidence in scientific research. *Social Studies of Science* 49: 732–757.

Mayernik, Matthew. 2020. Metadata. In *Encyclopedia of knowledge organization*, edited by Birger Hjørland and Claudio Gnoli. https://www.isko.org/cyclo/metadata#col.

Mayr, Ernst. 1988. *Towards a New Philosophy of Biology: The Observations of an Evolutionist*. Cambridge, MA: Harvard University Press.

McDonough, Jerome, Matthew Kirschenbaum, Doug Reside, Neil Fraistat, and Dennis Jerz. 2010. Twisty little passages almost all alike: Applying the FRBR model to a classic computer game. *Digital Humanities Quarterly* 4(2). http://www.digitalhumanities.org/dhq/vol/4/2/000089/000089.html.

McGann, Jerome. 1983. *A Critique of Modern Textual Criticism*. Chicago: University of Chicago Press.

McGann, Jerome. 2001. *Radiant Textuality: Literary Studies After the World Wide Web*. New York: Palgrave Macmillan.

McKenzie, D. F. 1999. *Bibliography and the Sociology of Texts*. Cambridge, UK: Cambridge University Press.

McNeil, Heather. 1994. Archival theory and practice: Between two paradigms. *Archivaria* 37: 6–20.

McPherson, Tara. 2018. *Feminist in a Software Lab: Difference + Design*. Cambridge, MA: Harvard University Press.

Melo, Marijel, and Laura March. 2022. By the book: A pedagogy of authentic learning experiences for emerging makerspace information professionals. *Journal of Education for Library and Information Science*.

Miller, Carolyn. 1984. Genre as social action. *Quarterly Journal of Speech* 70: 151–167.

Millerand, Florence, and Geoffrey Bowker. 2009. Metadata standards: Trajectories and enactment in the life of an ontology. In *Standards and Their Stories: How Quantifying, Classifying, and Formalizing Practices Shape Everyday Life*, edited by Susan Leigh Star and Martha Lampland. Ithaca, NY: Cornell University Press.

Mills, Jack. 1960. *A Modern Outline of Library Classification*. London: Chapman and Hall.

Moallem, Jon. 2017. Neanderthals were people, too. *New York Times Magazine*, January 11, 2017. https://www.nytimes.com/2017/01/11/magazine/neanderthals-were-people-too.html.

Mol, Annemarie. 2002. *The Body Multiple: Ontology in Medical Practice*. Durham, NC: Duke University Press.

Montoya, Robert and Katherine Morrison. 2019. Document and data continuity at the Glenn A. Black Laboratory of Archaeology. *Journal of Documentation* 75(5): 1035–1055.

Muller, Michael, Ingrid Lange, Dakuo Wang, David Piorkowski, Jason Tsay, Q. Vera Liao, Casey Dugan, and Thomas Erickson. 2019. How data science workers work with data: Discovery, capture, curation, design, creation. In *Proceedings of the 2019 CHI Conference on Human Factors in Computing Systems (CHI '19)*, 1–15. New York: Association for Computing Machinery.

National Information Standards Organization (NISO). 2011. ANSI/NISO Z39.50–2005 (R2010). *Guidelines for the Construction, Format, and Management of Monolingual Controlled Vocabularies.* https://groups.niso.org/apps/group_public/download.php/12591/z39-19-2005r2010.pdf.

Noble, Safiya. 2018. *Algorithms of Oppression: How Search Engines Reinforce Racism.* New York: New York University Press.

Nobles, Melissa. 2000. *Shades of Citizenship: Race and the Census in Modern Politics.* Stanford, CA: Stanford University Press.

Odom, William, John Zimmerman, and Jodi Forlizzi. 2014. Placelessness, spacelessness, and formlessness: Experiential qualities of virtual possessions. In *Proceedings of the 2014 ACM Conference on Designing Interactive Systems (DIS '14):* 985–994. New York: Association for Computing Machinery.

Olson, Hope. 2002. *The Power to Name: Locating the Limits of Subject Representation in Libraries.* Dordrecht, Netherlands: Kluwer Academic.

Olson, Hope, and Rose Schlegl. 2001. Standardization, objectivity, and user focus: A meta-analysis of subject access critiques. *Cataloging and Classification Quarterly* 32(2): 61–80.

Olson, Hope, and Dieter Wolfram. 2008. Syntagmatic relationships and indexing consistency on a larger scale. *Journal of Documentation* 64(4): 602–615.

Open Science Collaboration. 2015. Estimating the reproducibility of psychological science. *Science* 349(6251): aac4716.

Park, Jung-Ran. 2009. Metadata quality in digital repositories: A survey of state of the art. *Cataloging and Classification Quarterly* 47(3–4): 213–228.

Pawley, Christine. 2006. Unequal legacies: Race and multiculturalism in the LIS curriculum. *Library Quarterly* 76(2): 149–68.

PBS Digital Studios. 2018. *Why Do We Say "Asian American" Not "Oriental"?* Origin of Everything series, season 1, episode 35. Aired July 10, 2018. https://www.pbs.org/video/why-do-we-say-asian-american-not-oriental-4mohsx/.

Peet, Lisa. 2016. Library of Congress drops illegal alien subject heading, provokes backlash legislation. *Library Journal,* June 13, 2016. https://www.libraryjournal.com/?detailStory=library-of-congress-drops-illegal-alien-subject-heading-provokes-backlash-legislation.

Peng, Roger D. 2011. Reproducible research in computational science. *Science* 334(6060): 1226–1227.

Pine, Kathleen H., and Max Liboiron. 2015. The politics of measurement and action. In *Proceedings of the 33rd Annual ACM Conference on Human Factors in Computing Systems (CHI '15),* 3147–3156. New York: Association for Computing Machinery.

Plantin, Jean Christophe, Carl Lagoze, Paul Edwards, and Christian Sandvig. 2018. Infrastructure studies meet platform studies in the age of Google and Facebook. *New Media and Society* 20(1): 293–310.

Pollack, Andrew. 2016. Mylan tries again to quell pricing outrage by offering generic EpiPen. *New York Times*, August 29, 2016. https://www.nytimes.com/2016/08/30/business/mylan-generic-epipen.html.

Post, Colin, Patrick Golden, and Ryan Shaw. 2018. Never the same stream: Netomat, XLink, and metaphors of web documents. In *Proceedings of the ACM Symposium on Document Engineering 2018 (DocEng '18)*, Article 13, 1–10.

Powell, Andy, et al. 2007. Dublin Core abstract model. https://www.dublincore.org/specifications/dublin-core/abstract-model/.

Prewitt, Kenneth. 2013. *What Is "Your" Race? The Census and Our Flawed Efforts to Classify Americans*. Princeton, NJ: Princeton University Press.

Ranganathan, S. R. 1957. *Prolegomena to Library Classification*. 2nd ed. London: Association of Assistant Librarians.

Ranganathan, S. R. 1959. *Elements of Library Classification*. London: Association of Assistant Librarians.

Rao, Tejal. 2017. One cookie, two versions: Why Girl Scout S'mores won't all be the same. *New York Times*, January 13, 2017. https://www.nytimes.com/2017/01/13/dining/smores-girl-scout-cookies.html.

Rayward, Boyd. 1994. Visions of Xanadu: Paul Otlet and hypertext. *Journal of the American Society for Information Science* 45(4): 235–250.

Reddy, Michael. 1979. The conduit metaphor: A case of frame conflict in our language about language. In *Metaphor and Thought*, edited by A. Ortony, 284–310. Cambridge, UK: Cambridge University Press.

Redström, Johan. 2008. RE: Definitions of use. *Design Studies* 29(4): 410–23.

Renear, Allen, and David Dubin. 2007. Three of the FRBR entities are roles, not types. In *Proceedings of the Annual Meeting of the American Society for Information Science and Technology 2007*, 44(1). https://doi.org/10.1002/meet.1450440248.

Ribes, David. 2019. How I learned what a domain was. In *Proceedings of the Association for Computing Machinery (ACM) on Human-Computer Interaction* 3(CSCW): Article 38.

Rich, Motoko. 2020. Japan loves robots, but getting them to do human work isn't easy. *New York Times*, January 1, 2020. https://www.nytimes.com/2019/12/31/world/asia/japan-robots-automation.html.

Richardson, Ernest Cushing. 1930. *Classification, Theoretical and Practical*. 3rd ed. Hamden, CT: Shoe String Press.

Rieder, Bernhard. 2020. *Engines of Order: A Mechanology of Algorithmic Techniques*. Amsterdam: Amsterdam University Press.

Rieh, Soo Young, Kevyn Collins-Thompson, Preben Hansen, and Hye-Jung Lee. 2016. Towards searching as a learning process: An overview of current perspectives and future directions. *Journal of Information Science* 42(1):19–34.

Roberts, Sarah. 2019. *Behind the Screen: Content Moderation in the Shadows of Social Media*. New Haven, CT: Yale University Press.

Rogers, Katie, Laura Jakes, and Ana Swanson. 2020. Trump defends using 'Chinese virus' label, ignoring growing criticism. *New York Times*, March 18, 2020. https://www.nytimes.com/2020/03/18/us/politics/china-virus.html.

Rooksby, John, Mattias Rost, Alistair Morrison, and Matthew Chalmers. 2014. Personal tracking as lived informatics. In *Proceedings of the 32nd Annual ACM Conference on Human Factors in Computing Systems (CHI '14)*, 1163–1172. New York: Association for Computing Machinery. http://doi.acm.org/10.1145/2556288.2557039.

Rosner, Daniela. 2012. The material practices of collaboration. In *Proceedings of ACM CSCW 2012*, 1155–1164. New York: Association for Computing Machinery.

Rosner, Daniela. 2018. *Critical Fabulations*. Cambridge, MA: MIT Press.

Rothstein, Edward. 2006. The anti-Semitic hoax that refuses to die. *New York Times* April 21, 2006. https://www.nytimes.com/2006/04/21/arts/design/the-antisemitic-hoax-that-refuses-to-die.html.

Said, Edward. 1978. *Orientalism*. New York: Pantheon Books.

Sanger-Katz, Margot. 2016. Is terrorism getting worse? In the West, yes. In the world, no. *New York Times*, August 16, 2016. https://www.nytimes.com/2016/08/16/upshot/is-terrorism-getting-worse-in-the-west-yes-in-the-world-no.html.

Santora, Marc S. 2018. Both sides claim victory in Macedonia's vote on changing its name. *New York Times*, September 30, 2018. https://www.nytimes.com/2018/09/30/world/europe/macedonia-greece-referendum.html.

Saracevic, Tefko. 1975. Relevance: A review of and a framework for thinking about the notion in information science. *Journal of the American Society for Information Science and Technology* 26(6): 321–343.

Saracevic, Tefko. 2007a. Relevance: A review of the literature and a framework for thinking on the notion in information science. Part II: Nature and manifestations of relevance. *Journal of the American Society for Information Science and Technology* 58(2): 1915–1933.

Saracevic, Tefko. 2007b. Relevance: A review of the literature and a framework for thinking on the notion in information science. Part III: Behavior and effects of relevance. *Journal of the American Society for Information Science and Technology* 58(3): 2126–2144.

Saracevic, Tefko. 2017. *The Notion of Relevance in Information Science. Everybody Knows What Relevance Is. But What Is It Really?* Synthesis Lectures on Information Concepts, Retrieval, and Services. Edited by Gary Marchionini. Kentfield, CA: Morgan & Claypool. https://doi.org/10.2200/S00723ED1V01Y201607ICR050.

Satariano, Adam. 2020. British grading debacle shows pitfalls of automating government. *New York Times*, August 20, 2020. https://www.nytimes.com/2020/08/20/world/europe/uk-england-grading-algorithm.html.

Sayers, W. C. 1915. *Canons of Classification*. London: Grafton.

Schmidt, Stefan. 2009. Shall we really do it again? The powerful concept of replication is neglected in the social sciences. *Review of General Psychology: Journal of Division 1 of the American Psychological Association* 13(2): 90–100.

Sebald, W. G. 1990. *Vertigo*. Translated by Michael Hulse. New York: New Directions.

Shavit, Ayelet, and James Griesemer. 2011. Transforming objects into data: How minute technicalities of recording "species location" entrench a basic challenge for biodiversity. In *Science in the Context of Application*, edited by M. Carrier and A. Nordmann, 169–193. Boston Studies in the Philosophy of Science, v. 274. Dordrecht, Netherlands: Springer.

Shaw, Ryan. 2019. The missing profession: toward an institution of critical technical practice. *Information Research* 24(4). http://www.informationr.net/ir/24-4/colis/colis1904.html.

Shetterly, Margot. 2016. *Hidden Figures: The American Dream and the Untold Story of the Black Women Mathematicians Who Helped Win the Space Race*. New York: William Morrow.

Shilton, Katie. 2013. Values levers: Building ethics into design. *Science, Technology, and Human Values* 38(3): 374–397.

Skouvig, Laura. 2020. The raw and the cooked. In *Forms of Knowledge: Developing the History of Knowledge*, edited by Johan Östling, David Larsson Heidenblad, and Anna Nilsson Hammar, 108–121. Lund, Sweden: Nordic Academic Press.

Simonite, Tom. 2019. Algorithms should have made courts more fair. What went wrong? *Wired*, September 5, 2019.

Skinner, Ginger. 2016. Can you get a cheaper EpiPen? *Consumer Reports*, August 11, 2016. https://www.consumerreports.org/drugs/can-you-get-a-cheaper-epipen/.

Smith, Helena. 2014. Alexander the Great claimed by both sides in the battle over name of Macedonia. *The Guardian*, November 10, 2014. https://www.theguardian.com/world/2014/nov/10/alexander-the-great-macedonia-greece.

Smith, Helena. 2019. Macedonia officially changes its name to North Macedonia. *The Guardian*, February 12, 2019. https://www.theguardian.com/world/2019/feb/12/nato-flag-raised-ahead-of-north-macedonias-prospective-accession.

Snyder, Jaime. 2017. Vernacular visualization practices in a citizen science project. In *Proceedings of the 2017 ACM Conference on Computer Supported Cooperative Work and Social Computing (CSCW '17)*, 2097–2111. New York: Association for Computing Machinery.

Soergel, Dagobert. 1974. *Indexing Languages and Thesauri: Construction and Maintenance*. Los Angeles: Melville.

Soto, Michael. 2016. *Measuring the Harlem Renaissance: The U.S. Census, African American Identity, and Literary Form*. Amherst: University of Massachusetts Press.

Southern Poverty Law Center. 2017. The miseducation of Dylann Roof. https://www.youtube.com/watch?v=qB6A45tA6mE.

Spärck-Jones, Karen, and Peter Willet, eds. 1997. *Readings in Information Retrieval*. New York: Morgan Kaufmann.

Srinivasan, Janaki, Megan Finn, and Morgan Ames. 2017. Information determinism: Consequences of faith in information. *The Information Society* 33(1): 13–22.

Star, Susan Leigh, and James Griesemer. 1989. Institutional ecology, "translations" and boundary objects: Amateurs and professionals in Berkeley's Museum of Vertebrate Zoology, 1907–1939. *Social Studies of Science* 19(3): 387–420.

Star, Susan Leigh, and Karen Ruhleder. 1996. Steps toward an ecology of infrastructure: Design and access for large information spaces. *Information Systems Research* 7(1): 111–134.

Stevens, Matt. 2020. How Asian-American leaders are grappling with xenophobia amid coronavirus. *New York Times*, March 29, 2020. https://www.nytimes.com/2020/03/29/us/politics/coronavirus-asian-americans.html.

Suchman, Lucy. 2002. Located accountabilities in technology production. *Scandinavian Journal of Information Systems* 14(2): Article 7.

Sutherland, Tonia. 2017. Archival amnesty: in search of Black American transitional and restorative justice. *Journal of Critical Library and Information Studies* 1(2). https://doi.org/10.24242/jclis.v1i2.42.

TallBear, Kim. 2013. Genetic articulations of indigeneity. *Social Studies of Science* 43(4): 509–533.

Tanselle, G. Thomas. 1990. *Textual Criticism and Scholarly Editing*. Charlottesville: University of Virginia Press.

Tanweer, Anissa, Brittany Fiore-Gartland, and Cecilia Aragon. 2016. Impediment to insight to innovation: Understanding data assemblages through the breakdown–repair process. *Information, Communication & Society* 19(6): 736–752.

Taub, Amanda, and Max Fisher. 2018. Where countries are tinderboxes and Facebook is a match. *New York Times*, April 21, 2018. https://www.nytimes.com/2018/04/21/world/asia/facebook-sri-lanka-riots.html.

Tavernise, Sabrina, and Richard A. Oppel, Jr. 2020. Spit on, yelled at, attacked: Chinese Americans fear for their safety. *New York Times*, March 23, 2020. https://www.nytimes.com/2020/03/23/us/chinese-coronavirus-racist-attacks.html.

Taylor, Derrick Bryson, and Christine Hauser. 2021. What to know about the Atlanta spa shootings. *New York Times*, March 17, 2021. https://www.nytimes.com/2021/03/17/us/atlanta-spa-shootings.html.

Tennis, Joseph. 2012. The strange case of eugenics: A subject's ontogeny in a long-lived classification scheme and the question of collocative integrity. *Journal of the American Society for Information Science and Technology* 63(7): 1350–1359.

Tenopir, Carol, Suzie Allard, Kimberly Douglass, Arsev Umur Aydinoglu, Lei Wu, Eleanor Read, Maribeth Manoff, and Mike Frame. 2011. Data sharing by scientists: Practices and perceptions. *PLoS One* 6(6): e21101.

Tenopir, Carol, Elizabeth D. Dalton, Suzie Allard, Mike Frame, Ivanka Pjesivac, Ben Birch, Danielle Pollock, and Kristina Dorsett. 2015. Changes in data sharing and data reuse practices and perceptions among scientists worldwide. *PLoS One* 10(8): e0134826.

Thomas, Neal. 2018. *Becoming-Social in a Networked Age*. New York: Routledge.

Thomer, Andrea and Karen Wickett. 2020. Relational data paradigms: What do we learn by taking the materiality of databases seriously? *Big Data and Society* 7(1).

Thylstrup, Nanna Bonde. 2018. *The Politics of Mass Digitization*. Cambridge, MA: MIT Press.

Tolmie, Peter, Andy Crabtree, Tom Rodden, James Colley, and Ewa Luger. 2016. "This has to be the cats": Personal data legibility in networked sensing systems. In *Proceedings of the 19th ACM Conference on Computer-Supported Cooperative Work & Social Computing (CSCW '16)*, 491–502. New York: Association for Computing Machinery.

Trace, Ciaran. 2017. Phenomenology, experience, and the essence of documents as objects. *Information Research* 22(1). http://www.informationr.net/ir/22-1/colis/colis1630.html.

Tripodi, Francesca. 2022. *The Propagandists' Playbook: How Conservative Elites Manipulate Search and Threaten Democracy*. New Haven, CT: Yale University Press.

Tsing, Anna Loewenhaupt. 2015. *The Mushroom at the End of the World: On the Possibility of Life in Capitalist Ruins*. Princeton, NJ: Princeton University Press.

Turnbaugh, P., et al. 2007. The Human Microbiome Project. *Nature* 449: 804–810.

Turner, Deborah. 2010. Orally based information. *Journal of Documentation* 66(3): 370–383.

United States, Office of Congresswoman Grace Meng. Meng bill to remove the term "Oriental" from U.S. law signed by President Obama. Press release, May 20, 2016. https://meng.house.gov/media-center/press-releases/meng-bill-to-remove-the-term-oriental-from-us-law-signed-by-president.

United States Census Bureau. 2017. *2015 National Content Test Race and Ethnicity Analysis Report. A New design for the 21st Century*. https://www2.census.gov/programs-surveys/decennial/2020/program-management/final-analysis-reports/2015nct-race-ethnicity-analysis.pdf.

United States Census Bureau. n.d. Research to improve data on race and ethnicity. Accessed January 3, 2020. https://www.census.gov/about/our-research/race-ethnicity.html.

Urban, Richard. 2014. The 1:1 principle in the age of linked data. In *Proceedings of the International Conference on Dublin Core and Metadata Applications 2014*, 119–128.

Verbeek, Peter-Paul. 2005. *What Things Do: Philosophical Reflections on Technology, Agency, and Design*. University Park, PA: Penn State University Press.

Vertesi, Janet, and Paul Dourish. 2011. The value of data: considering the context of production in data economies. In *Proceedings of the ACM Conference on Computer Supported Cooperative Work 2011 (CSCW '11)*, 533–542. New York: Association for Computing Machinery.

Vesna, Victoria, editor. 2007. *Database Aesthetics: Art in the Age of Information Overflow*. Minneapolis: University of Minnesota Press.

Vickery, Brian. 1960a. *Faceted Classification: A Guide to the Construction and Usage of Special Schemes*. London: Aslib.

Vickery, Brian. 1960b. Thesaurus: A new word in documentation. *Journal of Documentation* 16(4): 181–189.

Vickery, Brian. 1975. *Classification and Indexing in the Sciences*. 3rd ed. London: Butterworths.

Wallis, Jillian C., Elizabeth Rolando, and Christine L. Borgman. 2013. If we share data, will anyone use them? Data sharing and reuse in the long tail of science and technology. *PloS One* 8(7): e67332.

Walsh, Fergus. 2020. AI "outperforms" doctors diagnosing breast cancer. *BBC*, January 2, 2020. https://www.bbc.com/news/health-50857759.

Wang, Hansi Lo. 2018. 2020 Census to keep racial: Ethnic categories used in 2010. *NPR*, January 26, 2018. https://www.npr.org/2018/01/26/580865378/census-request-suggests-no-race-ethnicity-data-changes-in-2020-experts-say.

Warner, Julian. 2009. *Human Information Retrieval*. Cambridge, MA: MIT Press.

Weagley, Julie, Ellen Gelches, and Jung-Ran Park. 2010. Interoperability and metadata quality in digital video repositories: a study of Dublin Core. *Journal of Library Metadata* 10(1): 37–57.

Weschler, Lawrence. 1996. *Mr. Wilson's Cabinet of Wonder: Pronged Ants, Horned Humans, Mice on Toast, and Other Marvels of Jurassic Technology*. New York: Vintage Books.

Wetzel, Linda. 2018. Types and tokens. In *The Stanford Encyclopedia of Philosophy* (Fall 2018 ed.), edited by Edward N. Zalta. https://plato.stanford.edu/archives/fall2018/entries/types-tokens/.

White, Kelvin L. 2017. Race and culture: an ethnic studies approach to archives and recordkeeping research in the United States. In *Research in the Archival Multiverse*, edited by Anne J. Gilliland, Sue McKemmish, and Andrew J. Lau, 352–381. Victoria, Australia: Monash University Publishing.

White, Ryen, and Resa Roth. 2009. *Exploratory Search: Beyond the Query-Response Paradigm*. Synthesis Lectures on Information Concepts, Retrieval and Services, edited by Gary Marchionini. Kentfield, CA: Morgan & Claypool.

Whittaker, Steve, and Julia Hirschberg. 2001. The character, value, and management of personal paper archives. *ACM Transactions on Computer-Human Interaction* 8(2): 150–170.

Wiegand, Wayne. 2015. *Part of Our Lives: A People's History of the American Library*. Oxford, UK: Oxford University Press.

Wilson, Patrick. 1968. *Two Kinds of Power: An Essay on Bibliographic Control*. Berkeley: University of California Press.

Wilson, Patrick. 1983. *Second-Hand Knowledge: An Inquiry into Cognitive Authority*. Westport, CT: Greenwood Press.

World Health Organization (WHO). 2015. Best practices for the naming of new human infectious diseases. https://apps.who.int/iris/bitstream/handle/10665/163636/WHO_HSE_FOS_15.1_eng.pdf.

Yang, Jia Lynn. 2020. When Asian Americans have to prove we belong. *New York Times*, April 10, 2020. https://www.nytimes.com/2020/04/10/sunday-review/coronavirus-asian-racism.html.

Yates, JoAnne. 1989. *Control through Communication: The Rise of System in American Management*. Baltimore: Johns Hopkins University Press.

Zeng, Marcia Lei, and Lois Mai Chan. 2010. Semantic interoperability. In *Encyclopedia of Library and Information Sciences*, 3rd ed., edited by Marcia Bates and Mary Niles Maack. Boca Raton, FL: CRC Press.

INDEX

Page numbers in italic indicate a figure and page numbers in bold indicate a table on the corresponding page.

Aboutness, in classification systems.
  See also Cataloguing; Indexing
  centrality of, 244n5
  concept-label relationships and, 178–183, 186–189
  definition of, 26, 84–85
  as "epistemic potential" of work, 271n27
  interpretive flexibility in, 219–225, **224**
  relevance versus, 48
  search engine approximations of, 272n30
  textual "clues" as proxies of, 271n29
  value of human judgment in, 225–230, 270n25
Abstraction, levels of, 61–64, 210, 223, 235–236
Abstract unity, 69
Access points, 85, 252n21
Accidental characteristics, 65–68, 75–76
Accountability, in information design, 29–31
Acker, Amelia, 257n6
Actor-network theory, 255n32
Adams, Richard, 246n6
Adler, Melissa, 265n16
Adventure essays
  on equivalence, 61–81
  on interoperability, 95–114
  on labels, 159–175
  on locality, 191–210
  on objectivity, 33–45
  on serendipity, 13–25
  structure of, 4–5
  on taxonomy, 129–135
Aesthetic force, data as, 245n12
African American Lives (TV series), 206
Agential realism, theory of, 269n15
Agre, Phil, 270n25
Alexander the Great, Macedonia naming dispute and, 164
Algorithms, 2
  decisions rendered by, 230
  judicial risk-assessment, 244n9, 245n11
  machine learning, 184
  minorities disadvantaged by, 244n9, 272n30
Algorithms of Oppression (Noble), 244n9, 272n30
Ali, Mohammad, 163
Aliasing, unlimited, 123–124
Alien, penumbral, umbral, penumbral, alien (APUPA) pattern, 26
All Data Are Local (Loukissas), 266n1
Amazon
  automated techniques at, 17, 227–228
  data ambiguity in, 79–80
  datafication and, 92
  as dataset, 4, 23

Index

Ambiguity of data
  abstraction and, 64
  functional equivalence and, 64–65, 75–81, *79*, 89, 235–239, 242
  in Metadata Architectures projects, 58, 235–239
  pervasiveness of, 3, 7, 242
  race and, 198–199
  relevance and, 48, 51
American Japanese Problem, The (Gulick), 171–172, 188–189
Ames, Morgan, 245n14
Amoore, Louise, 260n10
Ancestry.com, 206–207
Andersen, Jack, 257n6, 272n32
Andersen, Kristina, 249n16
Andersen Bageri (Copenhagen), 103–104, *104*
Anderson, Margo, 125, 267n6
Angwin, Julia, 244n9
Annunciation, The (di Benozzo), taxonomic description of, 129–131, *131*
Anzaldúa, Gloria, 10, 11
APUPA (alien, penumbral, umbral, penumbral, alien) pattern, 26
Arbuckle, Benjamin, 117
Arcades Project (Benjamin), 24–25
Archives, information provision and management in, 262n26
Arguello, Jaime, 50–51
Argument, taxonomy as, 134–135, 138
Arithmetic mean, inherent uncertainty in, 38–39
Armed Conflict Location and Event Data Project (ACLED), 23–24, 244n3
Arrangements, classification schemes conceptualized as, 86
Art, description as
  artistry and values of data creators in, 239–242
  data collection and, 235–239
  loving art concept, 231–236

Atici, Levent, 248n12
Attributes, data, 147–148
Audience perspective, impact of locality on, 191–193, 207–210
Austlit catalog, 76–77, *77*
Authenticity standards, 105–109
Authorial intention, 74, 251n10
Automated data collection devices, 117–122

Balkans, Macedonia naming dispute in, 163–166, 174
Barad, Karen, 255n32, 269n15
Barbrow, Sarah, 117, 272n31
Bardzell, Jeff, 258n11
Bardzell, Shaowen, 258n11
Bates, Jo, 248n12
Bates, Marcia, 27, 252n16
Bazerman, Charles, 272n32
Belgian Congo, 165
Bell Curve, The (Herrnstein), 225, 226
Bender, Emily, 51, 54
Benjamin, Ruha, 245n11
Benjamin, Walter, 24
Berman, Sanford, 186–188
BERT (Google), 51
Best practices for data implementation, 122, 128, 205, 220, 225, 254n27, 257n7
Bias
  in data interpretation, 54
  inherent nature of, 27–28, 152, 189, 239–240, 267n8
  in LCSH subject headings, 186–187
  responses to, 245n11
Bibliographic classification
  functional equivalence in, 68–74, *70*, *71*
  historical ideals of, 152, 156, 261n18–19, 261n20
  in knowledge organization, 86–87, 253n22, 253n24

Index   303

Binder, Thomas, 267n9
Biological taxonomy, 151–152,
   261n16
Black Planet (dating site), race and
   ethnicity in, **200**, 201–202
Blair, Ann, 269n20
Bliss, Henry Evelyn, 261n18, 271n27
Boltanski, Luc, 157
Borderlands (Anzaldúa), 10
Bowker, Geoffrey, 28, 117, 151, 218,
   248n12, 269n18
Brantley, Ben, 251n8
Broussard, Meredith, 246n6, 255n31
Buckland, Michael, 83, 118, 252n16,
   265n16
Bullard, Julia, 255n31, 262n22
Burch, Traci, 197–198
Burroughs, William S., 263n4
Bush, George H. W., 166
Bush, George P., 166
Bush, George W., 166
Business communication, quantitative
   metrics in, 246n3
Butter, examples of functional equivalence in, 61–64, 68, 77, 79–80, *79*,
   250n1

Care, matters of, 258n12
Carlyle, Allyson, 257n7
Carter, Daniel, 255n31, 257n6
Cartwright, Nancy, 114
Cataloging Cultural Objects (CCO),
   251n14
Cataloguing
   assumptions and orientation of,
      84–85
   creation/modification dates in,
      270n24
   digitization, 90, 118, 252n21,
      254n30, 255n31
   functional equivalence in, 68–74,
      70, *71*
   indexing compared to, 265n10
   interpretive flexibility in, 219–225,
      224
   objectives of, 252n17
   obsolescence of subject headings in,
      118–119
   racist/colonialist bias in, 186–187,
      265n16
   Resource Description and Access,
      252n18
   standardization of, 86–87
   subject headings in, 186–187,
      219–225, **224**, 253n24, 265n16
      (*see also* Cataloguing)
   variations of resources in, 270n23
Categorization, cognitive. *See* Cognitive categorization
Census Bureau. *See* US Census, race
   and ethnicity in
Central Person Registry (CPR),
   268n13
Characteristics, data
   accidental versus essential, 65–68,
      75–76
   definition of, 147–148
Chatman, Elfreda, 256n2
Chicago Manual of Style, The, 171,
   264n7
Chinese virus, 173, 264n9
Cho, Hyerim, 249n14
Clarke, Rachel, 269n21
Classification, 260n12. *See also*
   Cataloguing; Documents; Indexing;
   Taxonomies
   aboutness in. *See* aboutness, in classification systems
   arrangements in, 86
   complacency toward information
      seeking in, 245n14
   data design and implementation in,
      87–93
   enumerative, 253n25, 259n8
   functional equivalence in.
      See equivalence

Classification (cont.)
　indexing compared to, 180–182
　knowledge organization in.
　　See knowledge organization
　positionality and orientation in interactions with, 18–24, *21*
　power and accountability in, 29–31
　precoordinate versus postcoordinate, 253n25
　reciprocal themes across multiple disciplines, 28–29
　serendipity constrained by, 18–24, *18*
　serendipity facilitated by, 24–25
　standardization of, 86–87
　synthetic, 259n8
Classification and Indexing in Science (Vickery), 253n23
Classification and Indexing in the Humanities (Langridge), 253n23
Classification and Indexing in the Social Sciences (Foskett), 253n23
Classificationists, 3–4
Classification Research Group (CRG), 253n23
Cleaning data, 118, 121
Cleverdon, Cyril, 49, 247n7
Cloud classification, standardization across locations, 216–218, *217*, 269n16
Cloud Ethics (Amoore), 260n10
Coca Cola varieties, functional equivalence of, 65–68, 75–76
Codebooks, 244n3
Cognitive authority, 246n1
Cognitive categorization
　distributed cognition, 269n15
　integration with systematic data collection, 215–216
　prototype effects in, 210–216
Collections. *See also* Organizing systems
　definition of, 3–4, 243n1
　unexpected data stories in, 13–16

Collectives, 81–83, 92
Collins, Henry, 114–115
Collocation, 72, 75
Colonialism, legacy of nationality and race from, 207–210
Color classification, 213–216
Communication, mathematical theory of, 52–54, 248nn10–11
Computer programming, status of, 250n17
Concept-label relationships. *See* Labels, relationship with concepts
Conceptual replication, 115
Concern, matters of, 258n12
Conduit metaphor, 52–53
Constraints, 148
　in cognitive categorization, 210–216
　on taxonomies, 154–158, 262n25
Context of observation, 55, 248n12
Controlled vocabulary. *See also* Taxonomies
　complexity of, 148
　distinctions between concepts and labels in, 178–182
　empirical outcomes of, 123–126
　NISO standard for, 178–179, 181
　problems with, 124–126, 267n6
　unlimited aliasing in, 123–124
Conventions, social
　in business communication, 246n3
　measurement and, 40–45, *41*
Costanza-Chock, Sasha, 29
Couldry, Nick, 254n29
COVID-19 pandemic
　A-level grading scandal and, 246n6
　naming of, 173
Cranfield tests of indexing devices, 227, 271n28
Creativity
　in cognitive categorization, 210–216
　in data collection, 154–158
Critical Fabulations (Rosner), 30
Critical race theory (CRT), 270n25

Index   305

Cruising the Library (Adler), 265n16
Culture, sciences/humanities, 45–47
Curated folksonomy, 262n22
Cutter, Charles, 252n17

Dasgupta, Sayamindu, 249n16
Daston, Lorraine, 54, 216, 218
Data. *See also* Human-information
  interaction
  ambiguity of. *See* ambiguity of data
  bias in, 27–28, 54, 152, 186–187,
    189, 239–240, 245n11, 267n8
  characteristics of, 147
  cleaning, 118, 121
  collection of. *See* data collection
  constraints on, 148, 154–157, 220,
    242
  cooking, 244n7
  criticism of, 126–128, 258nn11–12
  definition of, 147–148, 248n11
  design and implementation of,
    87–93
  implementation of, 122, 205, 220,
    225, 254n27, 257n7
  as instrument of subjugation, 29
  "intuitive," 267n8
  nature of, 2–4
  ontological aspect of, 81–83, 93,
    255n33
  popular rhetoric around, 2–3
  quality assessment of, 269n21
  reuse of, 120, 258n9
  structured/unstructured, 147–148
  things as, 4
Databases
  datasets compared to, 254n28
  differences between, 256n4
  digital, 87
  Global Terrorism Database, 23
Data collection
  as art, 235–239
  as assertion of values, 219–225, **224**
  automated, 117–122

creativity in, 154–158
human work of, 57–59, 249n16–17
integration with cognitive categorization, 215–216
reproducibility crisis in, 115–116
standardization across locations,
  216–219, *217*
value of human judgment
  in, 225–230, *228*, 270n25,
  272n31–32
Data colonialism, 254n29
Data craft, 257n6
Data Feminism (D'Ignazio and Klein),
  266n1
Datafication, 92
Data scientists, 2, 59, 182
Datasets
  databases compared to, 254n28
  definition of, 243n1
  design and implementation of, 4
  examples of, 4
  structured, 3
Data violence, 163, 263n2
Dating.dk, 208
Dating sites, race and ethnicity data in,
  198–207, **200**
da Vinci, Leonardo, 90
Day, Ron, 252n16
Deadnames, 263n2
Decentering of humans, 93, 249n13
de Heem, Jan Davidsz, 231, *232*
Democratic Republic of the Congo
  (DRC), 165
Demographic data. *See* Race and
  ethnicity data
Denmark, race and ethnicity in,
  207–210, 268nn13–14
Description, as art
  artistry and values of data creators
    in, 239–242
  data collection and, 235–239
  loving art concept, 231–236
Descriptors, 178–182

Development, of taxonomies, 150–154
Dewey Decimal Classification (DDC), 180, 270n25
Dharma Match (dating site), race and ethnicity data in, **200**
Di Benozzo, Alesso, 131
Digitization, impact of, 90, 118, 252n21, 254n30, 255n31
D'Ignazio, Catherine, 29, 248n12, 266n1
Disciplinary indexes, 86–87
Discipline of Organizing, The (Glushko), 243n1
Distributed cognition, 269n15
Diverse Book Finder, 270n25
Division, principles of
   multiple principles of division, 141–145, 259n8
   single principle of division, 132–134, 138, 259n5
DNA analysis, of race/ethnicity, 205–207, 268n12
Documentarity (Day), 252n16
Documents
   aboutness of. *See* aboutness, in classification systems
   cataloguing theory of. *See* cataloguing
   definition of, 83–84
   extraction of concepts from, 182–185
   knowledge organization approach to, 86–87
   oral, 251n5
Domain analysis, 26, 156
Domains, 28, 177, 181, 184, 214
Donovan, Joan, 257n6
Doty, Mark, 231, 233
Dourish, Paul, 255n31
Drabinski, Emily, 265n16
Drucker, Johanna, 244n6, 254n30
Dublin Core Metadata Initiative (DCMI), 88–91, 254n27
Duck Duck Go, 228–229, *228*

Duhigg, Charles, 251n13
Duh warrant, 93
Dunbar, Anthony, 270n25
Dupré, John, 261n17

Eika, Jonas, 208–210
Elements, data, 147–148
Emanuel African Methodist Episcopal (AME) shooting, 248n8
Engines of Order (Rieder), 253n25, 271n28
Enjoyment, understanding versus, 22
Ensmenger, Nathan, 250n17
Enumerative classification, 253n25, 259n8
EpiPen price controversy, 80–81, 251n13
Episemantics of aboutness, 271n27
Equivalence, 6. *See also* Documents
   abstraction, levels of, 61–64, 210, 223, 235–236
   accidental versus essential characteristics of, 65–68, 75–76
   ambiguity in, 64–65, 75–81, *79*, 89, 235–239, 242
   in controlled vocabularies, 124
   data design and implementation, 87–93
   disciplinary assumptions about information things, 83–87
   flawed data models of, 76–81
   of information objects, 68–74, *70, 71*
   ontological aspect of, 81–83, 93, 255n33
Ereshefsky Marc, 252n15
ERIC database, 87
Essential characteristics, 65–68, 75–76
Ethnicity. *See* Race and ethnicity data
European Network Against Racism (ENAR), 268n14
Evidence-based decisions, 126
Exact replication, 115, 121–122

Index

Exclusionary patterns. *See also* Social justice
　in naming actions, 170–175
　in power relation hierarchies, 130

Facebook
　connection to eruptions of violence, 112, 257n5
　as dataset, 4, 23
　moderation of, 111–114, 257n6
Faceted classificatory structure
　definition of, 137
　purpose of, 259n8
　thesauri compared to, 137
　unclear taxonomic principles in, 141–145, **143**
Fake news, 29, 239, 257n6
Feminist Data Manifest-No, 245n13
Feminist standpoint epistemology, 7–8, 243n1
File transfer protocols, 118
Finn, Megan, 245n14
Fisher, Max, 257n5
Flair, 111–114, 257n7
Florilegia, 269n20
Folksonomy, 155–156, 262n22
Ford, Henry, 221
Forlizzi, Jodi, 255n31
Former Yugoslav Republic of Macedonia, 164
Foskett, A. C., 186
Foskett, D. J., 253n23
Fox, Violet, 249n14
Frederiksen, Mette, 208
Freeman, Jo, 262n25
French bread label, variety of concepts described by, 99–102, *101*, *102*
Friedman, Batya, 267n9
Frohmann, Bernd, 272n32
Functional equivalence. *See* Equivalence
Functional Requirements for Bibliographic Records (FRBR), 72–73, 251n6, 251n9

Furnas, George, 122–133
Furner, Jonathan, 265n13, 270n25

Galison, Peter, 54
Gates, Henry Louis Jr., 206, 268n11
Gelches, Ellen, 269n21
Gender equality, naming actions and, 166–169, 186–187, 263n2, 265n16
Genre conventions, 15, 110, 229, 272n32
Genus:species relationships, 259n4
Gibson, Amelia, 256n2
Gilbert, Jeremy, 262n25
Girl Scout cookies, examples of functional equivalence in, 64–65, 75, 251n4
Gitelman, Lisa, 244n6
Global Terrorism Database (GTD), 23–24
Glushko, Robert, 243n1
Gluten-Free Singles (dating site), race and ethnicity in, **200**, 202
Goldberg, Whoopi, 206
Golden, Patrick, 261n21
Good, Nathan, 122
Goodwin, Charles, 116, 248n12
Goodwin's color chart, 117
Google
　BERT, 51
　as collection, 23
Gorichanaz, Tim, 252n16
Grading criteria
　allure of numbers in, 40
　harm resulting from, 246n6
　quantitative versus qualitative aspects of, 42–43
　social conventions in, 40–45, 246n6
Graphs, relationships in, 262n24
Greece, Macedonia naming dispute in, 163–166, 174
Griesemer, James, 117, 119, 248n12
Gulick, Sidney, 171–172, 188–189
Gursoy, Ayse, 255n31

Haider, Jutta, 31
Haimson, Oliver, 263n2
Hamilton, Mary, 167–169, 174
Hamlet (Shakespeare), equivalent editions of, 68–74, *70*, *71*
Haraway, Donna, 30, 255n32
Hargraves, Ian, 267n9
Hauser, Elliott, 271n27
Heda, William Claesz, 231
Hidden Figures (Shetterly), 250n17
Hierarchy, power relations, 129–135, *131*, 258n2–3
Hierarchy, taxonomic, 6, 129–140
  faceted classificatory structure in, 137, 141–145, **143**
  inheritance in, 132
  power relations compared to, 130–135
  single principle of division in, 132–134, 138, 259n5
  structural coherence of, 133–140
  taxonomic analysis of, 135–140, *139*
  thematic coherence of, 133
Hirshberg, Julia, 255n31
Hjorland, Birger, 26, 156–157, 271n27
Hochschild, Jennifer, 197–198
Hoffman, Anna Lauren, 245n13, 263n2
Hogan, Kristen, 9–10, 243n2
Houser, Heather, 245n12
Hulme, E. W., 184
Human-information interaction
  mental labor required by, 31, 245n15
  taxonomies in, 148–150
Humanities, culture of, 45–47
Human judgment
  context of observation in, 55, 248n12
  decentering of human in, 93, 249n13
  description as art in, 231–242
  interpretive flexibility in, 219–225, **224**
  minimization of, 220–221, 269n20
  in quantitative measurement, 47–51, 54–57
  reproducibility and, 111–114, 257nn7–8
  retrieval evaluation process, 49–51
  value of, 225–230, *228*, 270n25, 272n31–32
  in work of data collection, 57–59, 249n16–17
Hutchins, Ed, 269n15
Huvila, Isto, 249n16

Identity, labels and
  exclusionary patterns in, 170–175, 264n7
  personal names, 159–163, 263n1
  place names, 163–166
  titles, 166–169, 174
Illegal alien, in LCSH subject headings, 265n16
Inclusion, rhetoric of, 245n13, 262n62
Indeterminability of species, 261n16
Indexing, 178–182
  Cranfield tests, 227, 271n28
  disciplinary indexes, 86–87
  history of, 265n10
  languages for, 87, 253n24, 253n223–224
Indexing It All (Day), 252n16
Individual flair, 111–114, 257n7
Information, definition of, 248n11
Information determinism, 245n14
Information marginalization, 256n2
Information objects
  abstract essence of, 68–69
  functional equivalence of, 68–74, *70*, *71*
Information poverty, 256n2
Information science
  complacent attitude toward information seeking in, 245n14
  intellectual heritage of, 26–27

precision and recall in, 33–35, 47–48, 247n7
relevance of, 9–11
Information theory. *See* Mathematical theory of communication (MTC)
Infowhelm, 245n12
Infrastructure studies, 28, 244nn7–8
Ingold, Tim, 1
Inheritance, in hierarchical relations, 132
Insider's world view, 256n2
International Baccalaureate (IB) program, grading scandal in, 246n6
International Classification of Diseases (ICD), 88, 151, 218–219, 261n14
International Standards Organization (ISO), 182
Interoperability, 6, 95. *See also* Reproducibility
  authenticity/inauthenticity standards in, 105–109
  controlled vocabulary for, 99–103, 123–126
  data criticism and, 126–128, 258nn11–12
  data reuse, 120, 258n9
  definition of, 6, 95, 118
  empirical outcomes of, 122–126, 258n10
  forms of, 118
  individual flair in, 111–114, 257n7
  inherent flaws of technical devices for, 105–114
  semantic versus structural, 116–122
  of space/location, 119
  standardized replicable processes, limitations of, 95–99
  vocabulary for, 103–105
Interpretive flexibility. *See also* Human judgment
  as assertion of values, 219–225, **224**
  in Metadata Architectures projects, 235–239
  minimization of, 220–221, 269n20
  standards development and, 269n18
  value of, 225–230, *228*, 270n25, 272n31–32
"Intuitive" data, 267n8
Irani, Lilly, 254n26
Is a relationships, 259n4
Islamophobia, 209

Jackson, Steven J., 117, 272n31
Jacobi, Derek, 69
JafariNaimi, Nassim, 267n9
Japanese-Americans, internment of, 172
JDate (dating site), race and ethnicity data in, 199, **200**
Jewishness, in race data, 199, **200**, 267n7
Johns Hopkins Coronavirus Research Center database, 244n3
Johnson, Mark, 248n9
Joint exhaustivity, in taxonomic hierarchies, 134–135, 259n7
Jones, Quincy, 206
Judicial risk assessment, 246n5, 246n6

Kansa, Eric, 117
Kansa, Sarah, 117
Kelly, Diane, 247n7
Keywords, 122, 147, 156, 259n6
Kind:instance relationships, 259n4
Kirk, David, 255n31
Kjellman, Ulrika, 239–242
Klein, Lauren, 29, 248n12, 266n1
Knowledge, definition of, 84, 248n11
Knowledge organization. *See also* Documents
  assumptions and orientation of, 86–87
  complacent attitude toward information seeking in, 245n14
  intellectual heritage of, 26–27
  power and accountability in, 29–31

Knowledge organization (cont.)
    precoordinate versus postcoordinate, 253n25
    reciprocal themes across multiple disciplines, 28–29
    serendipity in, 26
    social practice shaped by, 28, 244n7
    standardization of, 86–87
Koller, Alexander, 51, 54
Kripke, Saul, 261n15
Kunisada, Utagawa, *139*
Kwasnik, Barbara, 258n2
Kyle, Barbara, 261n21

Labels, relationship with concepts, 6
    common-sense distinctions between, 176–178
    concept identification process, 182–185
    in controlled vocabularies, 178–182
    exclusionary patterns in, 170–175, 264n7
    personal names, 159–163, 263n1
    pervasiveness of, 163–166
    place names, 163–166
    reconfiguration of, 186–189
    titles, 166–169, 174
La Frontera (Anzaldúa). See *Borderlands* (Anzaldúa)
Lakoff, George, 248n9
Langridge, D. W., 253n23
Latham, Kiersten, 252n16
Latinx, Census vocabulary for, 124–126, 198–199, 201, 266n4
Latour, Bruno, 117, 248n12, 255n32, 258n12
Law, John, 255n32, 255n33
Leavis, F. R., 45–46, 59
LeDantec, Christopher, 267n9
Lee, Erika, 264n8
Lee, Jin Ha, 249n14, 269n21
Leslie, Sarah-Jane, 261n15
A-level grading scandal (UK), 246n6

Levine, Caroline, 258n3
Levy, David, 252n16
Librarianship, 26, 226
    conceptions of taxonomy within, 152
    social justice and, 30, 270n25
    social status of, 260n12
Libraries, as organizing systems. *See also* Classification
    bibliographic classification, 86–87, 152, 156, 253n22, 253n24
    cataloguing theory in. *See* cataloguing
    classification systems in, 180–181
    digitization of, 90, 118, 252n21, 254n30, 255n31
    functional equivalence of information objects in, 69–70, *70*, *71*, 251n6, 251n9
    information provision and management in, 262n26
    knowledge organization in. *See* knowledge organization
    positionality and orientation in interactions with, 18–20
    serendipity constrained by, 18–20, *18*
    serendipity facilitated by, 13–16, 24–25
Library of Congress Subject Headings (LCSH), 85, 88, 180
    interpretive flexibility in, 219–225, **224**
    racist/colonialist bias in, 186–187, 265n16
Light, Ann, 157
Lightweight ontologies, 182
Lin, Yu-Wei, 248n12
Lindley, Siân, 263n4
Linked Data, 181–182
Literary warrant, 184, 187, 261n18–19
Liu, Alan, 254n30
Locality, 6, 191

data creation as assertion of values, 219–225, **224**
interoperability of, 119
measurement and, 193–196
prototype effects in, 210–216
race and ethnicity in online dating sites, 198–207, **200**
race and ethnicity in US Census, 196–198, **197**, 267n6
standardization of data collection and, 216–219, *217*
understanding shaped by, 191–193, 207–210
Logical data structure, 147–148
López-Alt, J. Kenji, 66
Louisiana Literature festival (Denmark), 192, 266n2
Loukissas, Yanni, 248n12, 266n1
Lubetzky, Seymour, 69, 72, 74, 84, 85
Lupton, Deborah, 246n2

Macaluso, Benoit, 116
Macedonia naming dispute, 163–166, 174
Machine learning, 184, 271n29
Magic Machine workshops, 249n16
Making domain, 249n16
Malcolm X, 163
MARC format, 252n21
March, Laura, 249n16
Marginalized communities
impact of data stories on, 29
knowledge and, 243n1
Margrethe II of Denmark, 175
Maron, Bill, 271n29
Marshall, Brianna, 249n16
Marshall, Cathy, 263n4
Martin, John, 256n2
Match (dating site), race and ethnicity data in, **200**
Mathematical theory of communication (MTC), 52–54, 248nn10–11
Matters of care, 258n12

Matters of concern, 258n12
Mazmanian, Melissa, 255n31
McPherson. Tara, 262n24
Mean, arithmetic, 38–39
Measurement
advantages and disadvantages of, 43–45
allure of numbers in, 38–40
in business communication, 246n3
conduit metaphor, 52–53
harm resulting from, 246n6
human judgment and, 47–51, 54–57
impact of locality on, 193–196
inherent uncertainty of, 35–39, *37*, 246n4
mathematical theory of communication, 52–54, 248nn10–11
observation in, 55, 248n12
precision of, 33–35, 47–48, 246n7
recall and, 47–48, 246n7
social conventions and, 40–45, *41*
two-cultures thinking in, 45–47, 57
Medical Subject Headings (MeSH), 177, 178
Mejias, Ulises, 254n29
Melo, Marijel, 249n16
Meng, Grace, 170
Meritocracy, privilege and, 8
Metadata, 254n27, 269n21
definition of, 260n13
Dublin Core Metadata Initiative, 88–91, 254n27
Metadata Architectures projects, 57–59, 249n15
data variation and interpretive judgment in, 235–239
quality of data in, 272n3
Metaphor
as cognitive mechanism, 248n9
conduit, 52–54
Michaels, Sheila, 168–169
Miller, Carolyn, 272n32
Millerand, Florence, 269n18

Moderation, social media, 111–114, 257n6
Montoya, Robert, 254n28
Morrison, Katherine, 254n28
Mr Jelly's Business (Upfield), equivalence of editions of, 76–77, 77, 78
Mrs./Miss/Ms. titles, identity conveyed by, 166–169
Multiple principles of division, 259n8
Murray-Wakelin, Janette, 222
Museum of Jurassic Technology (Los Angeles), 244n2
Museum of Natural and Artificial Ephemerata (MNAE), 16, 244n2
Museum of Vertebrate Zoology (MVZ), 119
Museums, as organizing systems, 15
Mutual exclusivity, in taxonomic hierarchies, 135

Names. *See* Labels, relationship with concepts
Nathan, Lisa, 267n9
National Information Standards Organization (NISO), 178–179, 181, 185
Nationality, race and, 207–210, 268nn13–14
Natural language processing (NLP), 51
Netflix, as collection, 23
    data ambiguity in, 80
    datafication and, 92
    lack of human judgments in, 17
Neutrality, rejection of, 7
Noble, Safiya Umoja, 244n6, 244n9, 248n8, 272n30
North Macedonia, Macedonia naming dispute in, 163–166, 174
Notations, 181
Numbers, allure of, 38–40

Obama, Barack, 268n11
Objectivity, 6
    allure of numbers in, 38–40
    conduit metaphor, 52–53
    context of observation in, 55, 248n12
    history of, 54–55
    human judgment in, 47–51, 54–57
    mathematical theory of communication, 52–54, 248nn10–11
    in Metadata Architectures projects, 235–239
    precision of measurement devices and, 33–35, 47–48, 246n7
    recall and, 47–48, 246n7
    relevance and, 48–51
    retrieval evaluation, process of, 49–51
    social conventions and, 40–45, 41
    as that which speaks for itself, 54–57
    two-cultures thinking in, 45–47, 57
    uncertainty of measurement and, 35–39, 37, 246n4
    in work of data collection, 57–59, 249n16–17
Object-oriented ontology, 255n32
Odom, William, 255n31
Office of Management and Budget (OMB), 125, 266n4
OK Cupid (dating site), race and ethnicity data in, **200**
Olson, Hope, 265n16
One:one principle, 88–91
Online content moderation, 249n16
Online dating sites, race and ethnicity in, 198–207, **200**
Ontologies. *See also* Taxonomies
    definition of, 148, 260n12
    lightweight, 182
Oral documents, 251n5
Organizing systems. *See also* Equivalence
    complacent attitude toward information seeking in, 245n14
    definition of, 243n1

Index   313

human judgments and decisions
    involved in, 15–18
intellectual heritage of knowledge
    organization in, 26–27
positionality and orientation in
    interactions with, 18–24, *21*
power and accountability in design
    of, 29–31
precoordinate versus postcoordinate,
    253n25
reciprocal themes across multiple
    disciplines, 28–29
serendipity constrained by, 18–20,
    18
serendipity facilitated by, 13–18,
    24–25
Orientalism (Said), 188
Oriental label, exclusionary nature of,
    170–172, 264n7
Otlet, Paul, 30–31

Parameters, data, 148
Park, Jung-Ran, 269n21
Partial perspective, 29–30
Part:whole relationships, 259n4
Personal names, identity conveyed by,
    159–163, 263n1
Personal preferences
    in interactions with organizing sys-
        tems, 18–24, *21*
    universalization of, 18–20
Perspective
    impact of locality on, 191–193,
        207–210
    limited nature of, 8, 243n1
    partial, 29–30
Perti, Andrew, 249n14, 269n21
Philosophy of science, 151–152,
    255n32, 261n15
Place names, identity conveyed by,
    163–166
Policing, predictive, 245n11
Pollack, Andrew, 251n13

Polysemy, 99–102, *101*, *102*
Positionality, in interactions with orga-
    nizing systems, 18–24
Post, Colin, 261n21
Postcoordinate classification schemes,
    253n25
Postphenomenology, 255n32
Powell, Alison, 157
Power, in information design, 29–31
Power relations, hierarchical, 6,
    129–135, *131*, 258n2–3
Power to Name, The (Olson), 265n16
Precision
    of measurement, 33–35, 47–48,
        193–196, 246n7
    of vocabulary, 103–105
Precoordinate classification schemes,
    253n25
Preferred terms, 178–182
Prejudices and Antipathies (Berman), 186,
    188
Prewitt, Kenneth, 267n6
Privilege, perspectives from, 8
Properties, data, 147–148
Protocols of the Elders of Zion
    editions of, 221, 269n21
    subject headings assigned to,
        219–226, **224**
Prototype effects
    local conditions and, 213–216
    significance of, 210–213
Proxy data, limitations of, 228–229,
    228
PubMed database, 87, 177
Puig de la Bellacasa, María, 258n12
Putnam, Hilary, 261n15

Quantified Self, The (2016), 246n2
Quantitative measurement
    advantages and disadvantages of,
        43–45
    allure of numbers in, 38–40
    in business communication, 246n3

Quantitative measurement (cont.)
  conduit metaphor, 52–53
  harm resulting from, 246n6
  human judgment in, 47–51, 54–57
  inherent uncertainty of, 35–39, *37*, 246n4
  mathematical theory of communication, 52–54, 248nn10–11
  as motivation for personal self-tracking, 246n3
  observation in, 55, 248n12
  precision of, 33–35, 47–48, 246n7
  recall and, 47–48, 246n7
  relevance in, 48–51
  social conventions and, 40–45, *41*
  two-cultures thinking in, 45–47, 57

Race After Technology (Benjamin), 245n11
Race and ethnicity data, 3, 207–210
  advantages and disadvantages of, 266n3
  classificatory structures and, 28, 244n9, 246n6, 248n8
  DNA analysis of, 205–207, 268n12
  ideological assumptions of data creators in, 239–242
  information marginalization and, 256n2
  in making domain, 249n13
  naming actions and, 166–169, 170–175, 186–187, 263n2, 264n7, 265n16
  nationality and, 207–210, 268nn13–14
  in online dating sites, 198–207, **200**
  role of locality in understanding of, 191–193
  Swedish Institute for Race Biology case study, 239–242
  ubiquity of, 196
  in US Census, 124–126, 196–198, **197**, 267n6

Ranganathan, S. R., 26, 120–121, 156, 253n25, 257n7, 259n8, 261n20
Rankine, Claudia, 192, 207
Raw Can Cure Cancer (Murray-Wakelin), subject headings assigned to, 222–223, 224–225, 226
Rayward, Boyd, 31
Readings in Information Retrieval (Spärck-Jones and Peter Willet), 247n7
Recall, 47–48, 246n7
Recipes, standardized processes in, 95–99, 255n1
Reddy, Michael, 52–54
Redlining, 266n3
Redström, Johan, 262n22
Reflection essays
  on equivalence, 81–94
  on interoperability, 114–128
  on labels, 176–189
  on locality, 210–231
  on objectivity, 45–59
  on serendipity, 26–32
  structure of, 5
Relevance, 48–51
Replication
  conceptual, 115
  exact, 115, 121–122
  reproduction versus, 114
Reproducibility, 95, 122–123. *See also* Interoperability
  authenticity/inauthenticity standards in, 105–109
  concept of, 114–116
  data reuse and, 120, 258n9
  definition of, 95
  goals of, 114
  human judgment and, 111–114, 257nn7–8
  individual flair in, 111–114
  inherent flaws of technical devices for, 105–114
  replication versus, 114
  reproducibility crisis, 115–116

role of individual flair in, 257n7
standardized replicable processes, limitations of, 95–99
Republic of the Congo, 165
Resource Description and Access (RDA), 221, 252n18
Resource Description Framework (RDF) statements, 181–182
Retrieval evaluation, process of, 49–51
Reuse of data, 258n9
Ribes, David, 28
Rieder, Bernhard, 253n25, 271n28, 271n29
Rifbjerg, Synne, 192, 207
Roof, Dylann, 248n8
Rosner, Daniela, 30, 255n31
Routine checking, 114

Said, Edward, 188
Sanger-Katz, Margot, 256n4
Saracevic, Tefko, 48
Satariano, Adam, 246n6
Scalar, 262n24
Scales, precision of, 33–35, 246n4
Schemas
  definition of, 147–148
  encoded structure of, 147–148, 260n12
Schlegl, Rose, 265n16
Schmidt, Stefan, 114
Science, philosophy of, 151–152, 255n32, 261n15
Sciences, culture of, 45–47
Scientific visualization, 54–57
Scientific warrant, 184, 187
Scrolling Forward (Levy), 252n16
Search engines
  controlled vocabulary for, 99–103, 123–126, 148
  limitations of, 228–229, *228*, 230, 272n30
Sebald, W. G., 24

Second-Hand Knowledge (Wilson), 246n1
Sellen, Abigail, 255n31
Semantic interoperability, 116–122
Serendipity, 13–31
  complacent attitude toward information seeking in, 245n14
  concept of, 5
  experiences constrained by organizing systems, 18–20, *18*
  experiences facilitated by organizing systems, 13–18, 24–25
  human judgments and, 15–18
  in knowledge organization, 26–27
  positionality and orientation and, 18–24, *21*
  reciprocal themes across multiple disciplines, 28–29
Severe acute respiratory syndrome (SARS), 173
Shakespeare, William, 69, 70. See also *Hamlet* (Shakespeare), equivalent editions of
Shannon, Claude, 52, 248n11
Shavit, Ayelet, 119, 248n12
Shaw, Ryan, 157, 261n21
Shetterly, Margot, 250n17
Shilton, Katie, 267n9
Shinawatra, Yingluck, 106
Shklovski, Irina, 157
Sholler, Daniel, 257n6
Simple Knowledge Organization System (SKOS), 181–182
Single principle of division, 132–134, 138, 259n5
Skouvig, Laura, 244n7
Skyrim, interpretive diversity in dataset for, 236–238
"Small worlds," 256n2
Snow, C. P., 45–47, 59
Snyder, Jaime, 257n7
Social conventions, measurement and, 40–45, *41*, 246n6

Social identities, knowledge and, 8, 243n1
Social justice. *See also* Race and ethnicity data
 community-based archives and, 262n26
 information science and, 30
 naming actions and, 163, 166–175, 186–187, 263n2, 264n7, 265n16
Social media moderation, 111–114, 257n6
Sociotechnical infrastructure, classification schemes as, 28
Søe, Sille Obelitz, 257n6
Soergel, Dagobert, 178, 181, 183–185, 186
Sorting Things Out (Bowker and Star), 28
Soto, Mike, 267n6
Southern Poverty Law Center, 248n8
Space, interoperability of, 119
Spärck-Jones, Karen, 247n7
Species, indeterminability of, 261n16
Spotify, data ambiguity in, 80
Srinivasan, Janaki, 245n14
Standardization
 across locations, 216–219, *217*
 interpretive flexibility in, 269n18
Standardized replicable processes, limitations of, 95–99
Star, Susan Leigh, 28, 117, 218
Step counters, inherent uncertainty of, 35–38, *37*
Still Life with a Glass and Oysters (de Heem), 231, *232*
Still Life with Oysters, a Silver Tazza, and Glassware (Heda), 231, *233*
Still Life with Oysters and Lemon (Doty), 231–233
Strategy, 87–93
Structural coherence, of taxonomic hierarchy, 133–140

Structural constraints, on taxonomies, 154–158, 262n25
Structural interoperability, 118–119
Structured data, 148
Structured datasets, 3
Subject headings, 253n24. *See also* Cataloguing
 interpretive flexibility in, 219–225, **224**
 racist/colonialist bias in, 186–187, 265n16
Suchman, Lucy, 29–30, 249n16
Sundin, Olof, 31
Supermarkets, as organizing systems
 human judgments and decisions involved in, 16
 positionality and orientation in interactions with, 20–24, *21*
Swedish Institute for Race Biology, ideological assumptions in, 239–242
Synonymy, 99–102, *101*, *102*
Synthetic classification, 259n8

TallBear, Kim, 268n10
Tanselle, G. Thomas, 251n10
Taub, Amanda, 257n5
Taxonomies
 advantages of, 140–141
 argumentation in, 134–135, 138
 biological, 151–152, 261n15
 creativity in, 154–158
 data concepts in, 147–148
 data user interaction with, 148–150
 definition of, 148, 260n12
 development of, 150–154
 hierarchical relations in. *See* hierarchy, taxonomic
 structural constraints on, 6, 154–158, 262n25
 unclear taxonomic principles, examples of, 141–145, **143**
 user-centered, 145–147, 262n26
 utility of, 157

Index 317

Tennis, Joseph, 271n27
Terms
  collection/concept identification process, 182–185
  preferred, 178–182
  unpreferred, 179–180
Terrorism, concept of, 23–24, 111, 256n4
Texas Book Festival, 266n2
Textual studies, 251n6, 251n11
Thai Delicious committee, authenticity standards set by, 105–109, 256n3
Thematic coherence, of taxonomic hierarchy, 133
Thesauri, 137, 178
Thingness
  abstract essence of, 68–69
  accidental versus essential characteristics of, 65–68, 75–76
  ambiguity in. *See* ambiguity of data
  in data design and implementation, 87–93
  disciplinary assumptions about, 83–87
  functional equivalence of, 68–74
  levels of abstraction in, 64
  things as data, 4
Thomas, Neal, 270n26
Thomer, Andrea, 254n28
Thylstrup, Nanna Bonde, 254n30
Titles, identity and identification with, 166–169, 174
Tokens, 250n2
Tous Les Jours bakery, 100–102, *102*, 104
Trace, Ciaran, 252n16
Transfer protocols, 118
Travel, transport versus wayfaring modes of, 1–2
Tripodi, Francesca, 245n14
Trump, Donald, 173, 174, 238, 263n9, 264n9, 273n4
Turner, Deborah, 251n5

Twitter
  as collection, 23
  moderation of, 111–114
Types, 250n2

Uncertainty. *See also* Equivalence
  data criticism and, 126–128
  of measurement, 35–39, *37*, 246n4
  in NISO standard, 183
  pervasiveness of, 6, 33, 101, 239, 242
  relevance and, 48
  in US Census, 125
Understanding
  enjoyment versus, 22
  impact of positionality and orientation on, 18–24
United Kingdom, A-level grading scandal in, 246n6
Universality, rejection of, 7, 8
University of Copenhagen, 5, 13, 149–150, *149*, 208
University of Queensland, Austlit project, 76–77
Unlimited aliasing, 123–124
Unpreferred terms, 179–180
Unstructured data, 147
Urban, Richard, 89–91
US Census, race and ethnicity in
  controlled vocabulary and racial identifications, 124–126, 267n6
  local character of race in, 196–198, **197**
User-centered taxonomies, 145–147, 262n26
User tagging, 155–156, 262n22
User warrant, 184

Values, data, 148
Value-Sensitive Design, 267n9
Van House, Nancy, 29–30
Vernacular visualization, 257n7
Vertigo (Sebald), 24–25

Vickery, Brian, 178, 244n5, 253n23
Video Game Metadata Schema (VGMS), 249n14
Visual Resources Association (VRA), 251n14
Vocabulary
  controlled, 99–103, 123–126, 148, 178–182, 267n6 (*see also* taxonomies)
  polysemy and synonymy in, 99–102, *101, 102*
  precision of, 103–105
Vollmer, Joan, 263n4

Wakkary, Ron, 249n16
Warner, Julian, 31
Warrant, 93, 255n34, 261n18
  literary, 184, 187, 261n18–19
  scientific, 184, 187
  user, 184
Wayfaring, 1–2
Weagley, Julie, 269n21
Weaver, Vesla, 197–198
Weaver, Warren, 52, 248n11
Weinstein, Harvey, 226
We Need Diverse Books, 270n25
Weschler, Lawrence, 244n2
Wetzel, Linda, 250n2
White, Kelvin, 270n25
Whittaker, Steve, 255n31
Wickett, Karen, 254n28
Wikipedia, data creation in, 249n16
Willet, Peter, 247n7
Wilson, David, 244n2
Wilson, Patrick, 27, 246n1, 270n25
Winfrey, Oprah, 206
Wisdom, definition of, 248n11
Works, definition of, 69, 72, 251n6
World Health Organization (WHO), 173
World Wide Web Consortium (W3C)
  Semantic Web initiative, 181–182

Xerox, 29
XLink, 261n21

Yates, JoAnne, 246n3, 272n32
Yellow Peril, 171–172, 186–189
Yugoslavia, Macedonia naming dispute in, 163–166, 174

Zaire, 165
Zeffirelli, Franco, 73, 75
Zimmerman, John, 255n31